GEOTECHNICAL ENGINEERING

Kruger Brentt
Publishers

GEOTECHNICAL ENGINEERING

Alina M. Tibaldi
Dina Lopez
Scott Giorgis

Kruger Brentt
Publishers

2025

Kruger Brentt Publishers UK. LTD.
Company Number 9728962

Regd. Office: 68 St Margarets Road, Edgware, Middlesex HA8 9UU

ISBN: 978-1-78715-364-6

Disclaimer:

Every possible effort has been made to ensure that the information contained in this book is accurate at the time of going to press, and the publisher and author cannot accept responsibility for any errors or omissions, however caused. No responsibility for loss or damage occasioned to any person acting, or refraining from action, as a result of the material in this publication can be accepted by the editor, the publisher or the author. The Publisher is not associated with any product or vendor mentioned in the book. The contents of this work are intended to further general scientific research, understanding and discussion only. Readers should consult with a specialist where appropriate.

Every effort has been made to trace the owners of copyright material used in this book, if any. The author and the publisher will be grateful for any omission brought to their notice for acknowledgement in the future editions of the book.

For information on all our publications visit our website at http://krugerbrentt.com/

PREFACE

"Geotechnical Engineering" embarks on an insightful exploration of the evolving landscape of geotechnical engineering, offering readers a comprehensive understanding of the latest concepts, methodologies, and technologies shaping the field. This book serves as an essential resource for students, researchers, practitioners, and industry professionals seeking to stay abreast of the cutting-edge developments in geotechnical engineering.

Geotechnical engineering plays a critical role in civil infrastructure development, environmental protection, and natural hazard mitigation by studying the behavior of soils, rocks, and other geological materials and their interactions with structures and the environment. With rapid advancements in technology and the growing complexity of engineering projects, there is a pressing need to embrace emerging concepts and innovative approaches in geotechnical practice.

The primary objective of this book is to provide readers with a comprehensive overview of emerging concepts in geotechnical engineering, covering key topics such as advanced soil mechanics, numerical modeling, geotechnical instrumentation, remote sensing applications, ground improvement techniques, and sustainable geotechnics. Each chapter offers insights, theoretical frameworks, practical examples, and case studies to help readers understand the principles and applications of these emerging concepts.

Contributions from leading experts in the field explore various aspects of geotechnical engineering, including the integration of geosynthetics and recycled materials in construction, innovative foundation systems for tall buildings and offshore structures, geoenvironmental engineering for waste containment and remediation, and the application of artificial intelligence and machine learning in geotechnical analysis and design.

Special emphasis is placed on sustainability, resilience, and risk management in geotechnical practice, highlighting the importance of considering environmental, social, and economic factors in engineering decision-making. By embracing emerging concepts and technologies, geotechnical engineers can address complex challenges more effectively and contribute to the sustainable development of infrastructure and the built environment.

We extend our sincere gratitude to the contributors whose expertise and dedication have enriched this book. It is our hope that "Geotechnical Engineering" will serve as a valuable resource and inspiration for readers interested in exploring the frontiers of geotechnical practice and advancing the state of the art in engineering geology, soil mechanics, and foundation engineering.

Alina M. Tibaldi

Dina Lopez

Scott Giorgis

CONTENTS

01 GEOTECHNICAL ENGINEERING FROM A COMPREHENSIVE VIEWPOINT: RECENT INNOVATIONS IN DEVELOPING AREAS

1. GEOTECHNICAL ENGINEERING: RECENT ADVANCES

Geotechnical engineering is a field of civil engineering that focuses on the study of soils and structures, including those that are located on or within these soils. The study of soil mechanics enables us to comprehend the fundamental physical characteristics that influence how soils respond to external forces, such as structural loads or variations in water content. Soil mechanics primarily focuses on studying the deformation properties of soils in order to improve the accuracy of predicting failure loads. Thus, it is more practical and potentially more precise to conceptualize the issue as a soil-structure system consisting of a permeable soil medium (in the form of a soil profile with several layers) and a foundation or similar geotechnical engineering structure. Regarding the design of soil-structure systems, settlement and ultimate bearing capacity are the primary criteria that must be met in order to endure external static or dynamic loads. Once the stability criteria are established, it is anticipated that the soil-structure system will meet these requirements over the duration of the project. Geotechnical engineering encompasses various subfields that deal with practical issues of soils, rocks, slopes, foundations, walls, and other related matters. Although soil mechanics principles serve as a comprehensive guide, the approach to solving each problem varies. Engineers, who are like skilled chefs in this field, must use their expertise in mathematics to choose the most effective and desirable engineering solution. With the continuous advancement of technology, it is increasingly common for engineering solutions to align with developing technologies that can be integrated into geotechnical engineering research and practice. This allows for the development of highly efficient, feasible, and precise solutions. To accomplish this complex task, geotechnical engineers rely on three available tools: (i) experimental methods, (ii) numerical methods, and (iii) analytical approaches. This introductory chapter briefly discusses three developing specialties of geotechnical engineering that have had substantial advancements in the past few decades. The three topics covered are: (i) numerical approaches in geotechnical engineering, (ii) unsaturated soils, and (iii) offshore geotechnics. This text provides a concise overview of the primary focus of each discipline of study and highlights current advancements in each area.

1.1 Geotechnical Engineering: Numerical Methods

Numerical approaches offer approximate solutions to intricate engineering problems. Soils are materials that are composed of different components, have different properties in different directions, and consist of multiple phases. They also have a non-linear connection between stress and strain. Therefore, any geotechnical engineering problem that involves soils, either as a material or as the complete domain, presents a certain degree of complexity that requires strong numerical approaches to address the problem. The intricate nature of the phenomenon is frequently linked to the characteristics of the material or the conditions near its boundaries, or sometimes both. Given the complexity of these problems, it is clear that there are no analytical answers. Therefore, numerical approaches are the only viable alternative to provide approximate solutions with an acceptable margin of error. Several widely utilized numerical methods in contemporary practice include the finite element method (FEM), finite difference method (FDM), boundary element method (BEM), discrete element method (DEM), finite volume method (FVM), material point method (MPM), as well as hybrid methods. Choosing the appropriate approach might be challenging, but in general, the choice of numerical method depends on the specific problem at hand. The most suitable approach will be determined by the desired unknowns in the problem, as well as the essential details of which variables need to be computed, where they should be calculated, and the desired level of accuracy. Numerical approaches are commonly used to solve physical problems. These methods typically involve defining the problem in the spatial domain and, if the problem is time-dependent, also considering a temporal domain. Subsequently, the governing equations are formulated within these domains (or solely along the boundaries, as indicated by BEM), using the field variables that represent the unknowns in the problem. In geotechnical engineering problems, the primary field variables are often the displacements of the solid skeleton and the pore pressures in the voids of soils. These pore pressures can be either pore water pressures or pore air pressures, depending on whether the soil is represented as a two-phase or a three-phase medium, respectively. Next, the governing equations can be discretized within established domains or directly defined at specific locations within the soil material. The required boundary and initial conditions are then determined based on field variables (or degrees of freedom) and their derivatives with respect to Cartesian coordinates and/or time. Once the discretized versions of the governing equations are derived, they must be solved using a suitable numerical method for every load or time increment. Although this straightforward approach can yield a numerical solution to the problem, it is crucial to recognize that the accuracy of the final solution depends on the level of simplification applied at the beginning of the process. After calculating the field variables, the stresses (or any other stress-state variable) are determined at certain places in the soil elemental level, known as material points or convergence/integration points. This is done by employing constitutive models that utilize the strain field. Strains are determined using the kinematic strain-displacement relations. The constitutive law for the soil skeleton describes the link between effective stress and strain. The constitutive equation for the soil pores is Darcy's law, which governs the flow of pore fluid based on differences in total head between two places in the soil.

Recent challenges in geotechnical engineering, where numerical analysis plays a crucial role, include:

- Constitutive theories for transient soil response. Here, particularly the theoretical models developed for predicting the cyclic behavior of soft soils and impact-induced response of special soils and rocks entail certain issues. Such a behavior can be physically modeled on a cyclic triaxial apparatus and numerically simulated in [1] (see Figure 1).
- Large deformation or post-failure behavior of soft soils. A serious issue related to soft ground is "post-liquefaction behavior" as well as settlements and lateral spreading of loosely deposited saturated sands subject to seismic motions. Another one is the pile-driving problem or modeling the penetration of piles into soil deposits. This is a typical example of a large deformation and post-failure behavior, which may pose a challenge in terms of numerical modelling.
- Uncertainties in distributions/initial conditions of mechanical quantities of soils. Geotechnical problems always suffer from lack of knowledge on soil properties and initial conditions. Inadequate and insufficient information in geotechnical problems frequently lead to a discrepancy between prediction and measurement, which implies that the prediction in these problems is accompanied with several types of uncertainties. While deterministic approach has since been quite useful in providing an estimate solution provided that there is certain amount of data assimilation and inverse analyses back the approach, probabilistic approach is also promising. On the latter approach, stochastic analysis is a technique that considers the uncertainty as a probabilistic variable. It is employed to assess the possibility of failure or other limit states of soil-structure systems. Probabilistic approach is directly related to reliability analysis and performance-based design.

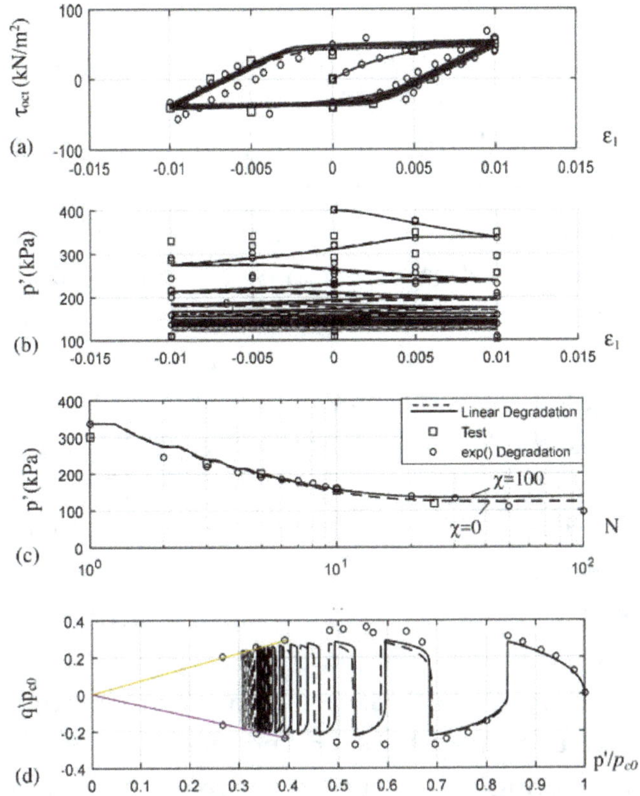

Figure 1. Two-way strain-controlled undrained triaxial test simulation of kaolin clay as compared with [2] Grenoble workshop tests and [3]. (a) Shear stress-axial strain relationship. (b) Decreasing mean effective stress with axial strain of 1 = ±1%. (c) Mean effective stress variation. (d) Stress path.

1.2 Unsaturated soils

Unsaturated soils are a growing area of interest in the discipline of geotechnical engineering. The behavior of unsaturated soils (USS) is a significant factor in various issues related to geotechnical engineering and geomechanics. The current geotechnical approach of treating unsaturated soil mechanical qualities as equivalent to fully saturated circumstances might lead to risky design outcomes, as it fails to account for the potential increase in soil water content. Conversely, when saturating natural soil samples in the laboratory, which typically mimic partially saturated conditions with water contents similar to those seen in the field, the soil becomes softer. This softening may result in excessive design measures. Due to the expected fluctuation in soil moisture levels in the field during the seasons, the dangers involved are increased. The advancements in the domain of "unsaturated soil mechanics" have not kept pace with those in "saturated soil mechanics," and the investigation of unsaturated soils has only lately become significant. Despite several studies undertaken over the past three decades, further experimental and theoretical research is required to fully comprehend the hydromechanical behavior of the USS. Before applying numerical approaches to solve a practical problem, it is necessary to calibrate the projected results of the established model via laboratory testing. This necessitates the incorporation of the model into a computer program employing numerical techniques specifically designed for a basic soil, utilizing USS parametric relationships within a multi-parametric framework. Therefore, the progress in the mechanics of the USS is directly linked to the advancements made in the preceding subject. One purpose is to eliminate the negative effects of treating USS as fully saturated, which can lead to unconservative design due to unpredictable behavior. However, the uneconomical design can be minimized by adopting a less cautious approach. The water content in unsaturated soils is crucial for assessing the hydromechanical response in relation to suction. The volume changes of USS under static loadings are greatly influenced by variations in their water content. However, this phenomenon is closely linked to the disparity in the pressures of pore water and pore air, which is referred to as "matric suction," s. The primary stress factor that determines the deformation features of USS is suction. As the matric suction grows, the undrained shear strength (USS) of the material also increases, along with its stiffness, yet the net mean stress remains constant. As for the stress state, while a single effective stress is sufficient to describe the stress-strain relationship of saturated soils, this is not the case for USS. Hence, effective stress equation needs to be modified in formulating a constitutive model for USS. This issue has been the main subject of debate among the researchers since the pioneering study of [4]. Bishop [5] is the first to propose a relation for effective stress that accounts for the air phase on the average stress acting on the solid skeleton. The "Bishop stress" is written as the modified effective stress:

$$\sigma'' = \sigma - \chi s \tag{1}$$

where χ is called the Bishop parameter which is a function of the water degree of saturation. While some researchers claim that there is a "smooth" transition to the new definition of the effective stress [6, 7, 8, 9], others say that there needs to be two independent stress variables, that is, the net stress, net or $\overline{\sigma}$, and matric suction to govern the response of USS [4, 10, 11]. Net stress is defined as:

$$\overline{\sigma} = \sigma - u_a \tag{2}$$

When creating a constitutive model for USS, it is crucial to consider the hydraulic hysteresis exhibited in the water retention behavior. The behavior described is readily apparent in a common connection known as the "soil-water characteristic curve (SWCC) or water retention curve (WRC)", regardless of the kind of soil. USS displays hysteresis during a drying-wetting cycle, which leads to variations in saturation and suction. Hence, it is imperative to include the hysteresis response in the mathematical representation of a constitutive model for Unmanned Surface Systems (USS). Table 1 provides a list of often occurring relationships. In this context, θ represents the volumetric water content, θs denotes the saturated water content as a percentage, θr refers to the residual volumetric water content, and Ψ represents the suction head. The parameters λ, a, m, and n are empirical constants associated with Ψ. Figure 2 displays the data acquired by [12] utilizing the HYPROP device, which is fitted to the curve by [13].

Table 1. Some commonly used SWCC relations.

Model	Relation
[14]	$\theta\Psi=\theta r+(\theta s-\theta r)(\alpha\Psi)-\lambda$
[13]	$\theta(\Psi) = \theta_r + \frac{\theta_s-\theta_r}{\{ln\,[e+(a\Psi)^n]^m\}}$
[15] (m = 1 - 1/n)	$\theta\Psi=\theta r+(\theta s-\theta r)[1+(\alpha\Psi)n]-m$
[15] (m, n independent)	$\theta(\Psi)=(\theta r+\theta s)[\ 1+(\alpha\Psi n)-m$

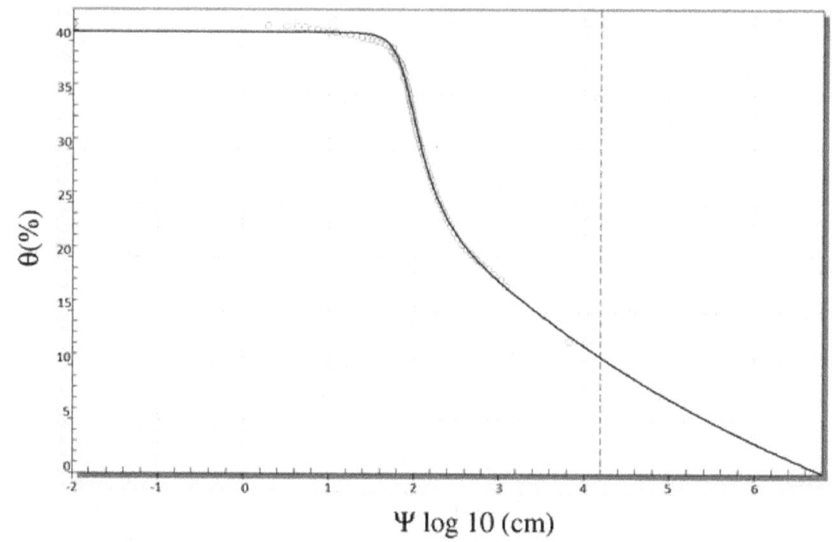

Figure 2. SWCC of Mersin silt fitted to Fredlund and Xing [13] (after [12]).

Recent elastoplastic models incorporating the hysteresis effect are distinguished in twofold: (i) models that account for hydraulic hysteresis through defining more yield surfaces [7, 16, 17] and (ii) models in which the effect of hydraulic hysteresis is included in the evolution of a so-called load collapse, (LC) curve [6, 18]. LC curve defines the evolution of "pre-consolidation pressure," p0 , of the soil with increasing plastic strains. Also, there are two other yield curves controlling the yielding of USS as matric suction changes called the suction increase (SI) and suction decrease (SD) yield curves (i.e., [16, 19]).

Barcelona Basic Model (BBM) [4] is considered to be the pioneering study that explicitly proposes an elastoplastic framework to understand the hydromechanical behavior of USS. The model is based upon the critical state theory with a classical plasticity framework employing a nonassociated flow rule. BBM adopts the yield surface of Modified Cam-Clay (MCC) model in the constant suction surface (Eq. (3)), whose size depends on suction (Figure 3). The yield function is written in triaxial stress components as:

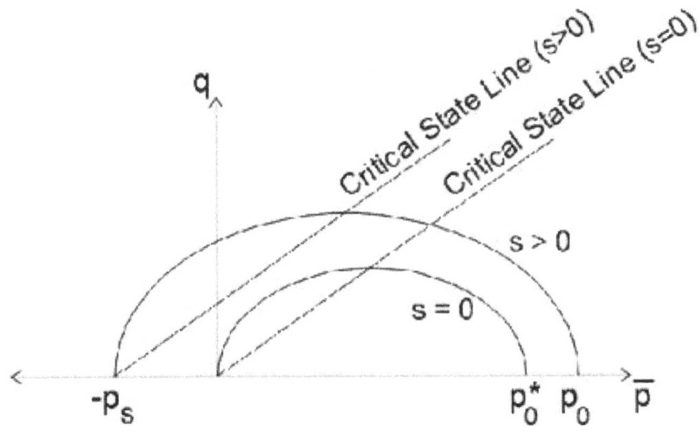

Figure 3. Dependence of the yield surface on suction.

$$q^2 - M^2(\overline{p} + p_s)(p_0 - \overline{p}) = 0 \tag{3}$$

As far as hydromechanical coupling, generally speaking, the yield criterion should consist of a hardening parameter that is a function of both suction and plastic strain, since stiffness and strength characteristics of unsaturated soils increase with increasing values of hardening parameters. Thus we have:

$$f(\overline{\sigma}, h) = 0 \tag{4}$$

$$h = h(\varepsilon^p, s) \tag{5}$$

where h is the "hardening parameter." In BBM per (3), h = p0 emphasizes that the consistency condition written for the yield surface must include suction as well as net mean stress. Consistency condition for the yield surface is then written as:

$$df = \frac{\partial f}{\partial \overline{\sigma}}^{T} d\overline{\sigma} + \frac{\partial f}{\partial p_0}\frac{\partial p_0}{\partial \varepsilon^p}^{T} d\varepsilon^p + \frac{\partial f}{\partial p_0}\frac{\partial p_0}{\partial s}^{T} ds + \frac{\partial f}{\partial p_s}\frac{\partial p_s}{\partial s} ds = 0 \tag{6}$$

While the formulation of the mechanical part of USS continues with the steps of classical plasticity starting from (Eq. (6)), the hydraulic part essentially starts off with the partial derivative of volumetric water content written as:

$$d\theta = S_r * dn + n * dS_r \tag{7}$$

where Sr is the degree of saturation and n is porosity. This equation can be expanded to yield a form employing matric suction and volumetric strain [20]. Derivation of Sr with respect to suction can then be obtained using the water retention behavior in terms of the soil-water characteristic curve. The final form of the coupling relation is obtained as:

$$d\bar{\sigma} = D_{hm}^{ep}\, d\varepsilon_{hm} \tag{8}$$

where Dep hm is the hydromechanical elastoplastic stiffness matrix. An example of predicted behavior of USS can be seen in Figure 4 as compared to the BBM.

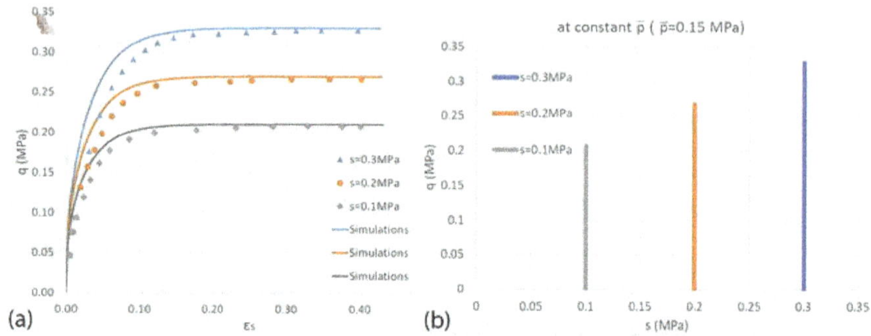

Figure 4. Prediction of hydromechanics of USS. (a) Verification with BBM [4] and (b) stress paths (after [20]).

Overall, the parameters that govern the USS response are intricate because of the particle, multiphase, and nonlinear characteristics of soils. The volume-saturation relationship and water retention behavior of unsaturated soils (USS) are more complex compared to saturated soils because of the presence of matric suction. Hydraulic hysteresis is a phenomenon observed in the water retention behavior of unsaturated soils, independent of their kind. Hence, a constitutive model designed to comprehend the hydromechanical behavior of unsaturated soils must consider these observable phenomena in specialized laboratory testing.

1.3 Offshore geotechnics

Offshore geotechnical engineering has experienced a surge in popularity during the past two decades. Offshore structures located on or near the seabed are crucial for energy production, defense, or transmission. The overall stability of the soil-structure system depends not only on the structural integrity during wave action, but also on the ability of the seabed soil to bear the stresses and pore pressure accumulation caused by waves and currents. The identification of geotechnical factors is crucial in determining the circumstances that can cause instability in offshore structural systems due to wave action. Hence, it is crucial to model the behavior of soil generated by waves using suitable theoretical frameworks. The advancements in this domain rely on three interconnected subtopics: (i) wave mechanics, (ii) structural mechanics, and (iii) soil mechanics. Therefore, the research of fluid-soil-structure interaction (FSSI) is essential for those involved in offshore geotechnics and requires careful attention. Furthermore, the research conducted in this area is also connected to the subjects addressed in the preceding sections, as numerical approaches are commonly utilized to offer approximate technical solutions to associated problems. Furthermore, although it is rare to find unsaturated soils in offshore environments (except for the air voids caused by sea shells), the principles of soil constitutive modeling related to unsaturated soils can indirectly be applied to understand the behavior of saturated soils found in seabed soils. Therefore, in this particular part of the introductory chapter, the emphasis is on the connections between seabed soils and their behavior when subjected to repeated wave loading, rather than on specific offshore constructions and the intricacies of their study. Those who are interested can consult the latest proceedings published by the International Society of Offshore and Polar Engineers (ISOPE) [21, 22, 23]. Figure 5 depicts the issue of wave-induced bottom erosion in an open area. Wave loading in this context causes a comparable variation in harmonic stress on a specific layer of the seafloor. The objective is to determine the displacements and pore pressures in the seabed caused by wave action. Undoubtedly, the true wave effect will have an irregular structure. However, this model adequately captures the system's response to wave-induced forces.

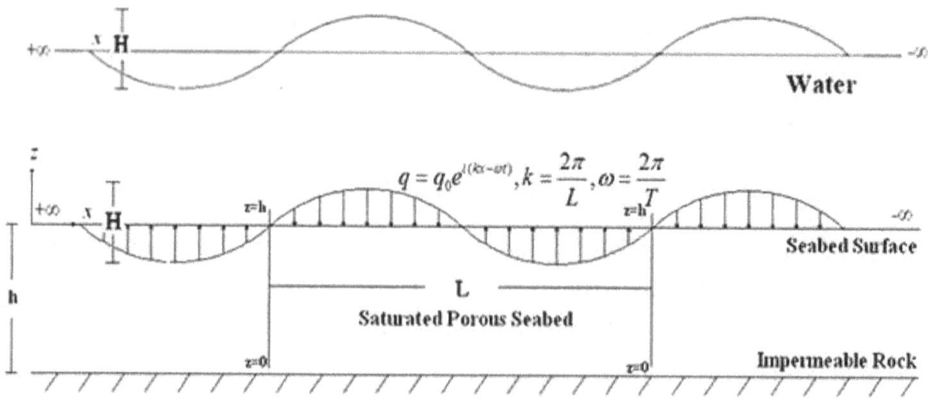

Figure 5. Free field wave-induced saturated porous seabed response.

The problem's solution, which involves analyzing and calculating the dynamic reaction of seabed soil, is presented in Figure 6a, b. The equations used to describe this response are those of linked flow and deformation in Biot poroelasticity [24, 25]. The corresponding references for Figure 6a and b are [26, 27]. Before utilizing any numerical techniques to create a related model, it is strongly advised that the researchers address the issue of the free field response. This will provide them with a more comprehensive grasp of the fundamental characteristics of a saturated porous seabed. Furthermore, this approach allows for the modeling of the wave-induced instantaneous liquefaction phenomenon found in the seabed, based on the mean effective stress (Sm) criterion [28] (Figure 6c). Readers who are interested in further exploring the subject matter in this sector can consult the extensive literature that is accessible on the issue, including references [28, 29, 30, 31].

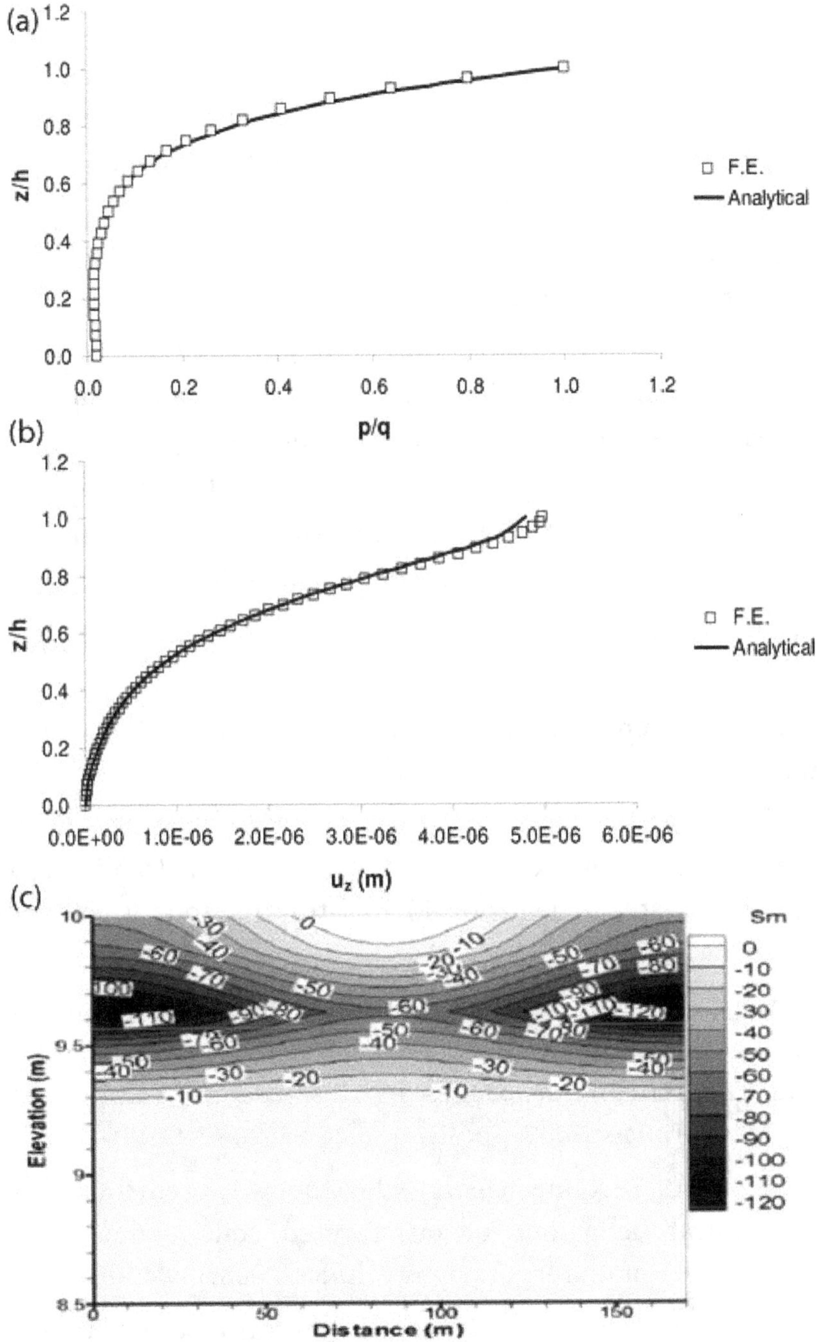

Figure 6. (a) Pore pressure, (b) displacement distribution and (c) liquefaction response within the seabed.

REFERENCES

1. Ülker MBC. A new hardening interpolation rule for the dynamic behavior of soils using generalized plasticity framework. In: ICSMGE 2017—19th International Conference on Soil Mechanics and Geotechnical Engineering; 2017
2. Kuntsche K. Tests on clay. In: Proc. International Workshop on Constitutive Behavior of Soils; Grenoble; 1982
3. Zienkiewicz OC, Leung KH, Pastor M. Simple model for transient soil loading in earthquake analysis. I. Basic model and its application. International Journal for Numerical and Analytical Methods in Geomechanics. 1985;9:453-476. DOI: 10.1002/nag.1610090505
4. Alonso EE, Gens A, Josa A. A constitutive model for partially saturated soils. Geotechnique. 1990;40:405-430
5. Bishop A. The principle of effective stress. Teknisk Ukeblad. 1959;39:859-863
6. Gallipoli D, Gens A, Sharma R, Vaunat J. An elasto-plastic model for unsaturated soil incorporating the effects of suction and degree of saturation on mechanical behavior. Géotechnique. 2003;53:123-136
7. Wheeler SJ, Sharma R, Buisson M. Coupling of hydraulic hysteresis and stress–strain behavior in unsaturated soils. Geotechnique. 2003;53:41-54
8. Jommi C. Remarks on the constitutive modelling of unsaturated soils. In: Tarantino A, Mancuso C, editors. Experimental Evidence and Theoretical Approaches in Unsaturated Soils. Rotterdam: Taylor & Francis; 2000. pp. 121-138. ISBN: 90-5809-186-4
9. Sun D, Sheng D, Xiang L, Sloan S. Elastoplastic prediction of hydro-mechanical behaviour of unsaturated soils under undrained conditions. Computers and Geotechnics. 2008;35:845-852
10. Fredlund D, Morgenstern N. Stress state variables for unsaturated soils. Journal of Geotechnical and Geoenvironmental Engineering. 1977;103:447-466

11. Fredlund DG, Pham HQ. Independent roles of the stress state variables on volume–mass constitutive relations. Theoretical and Numerical Unsaturated Soil Mechanics. Berlin, Heidelberg: Springer; 2007:37-44. DOI: 10.1007/3-540-69876-0_5

12. Eyupgiller M, Ulker M. Determination of the relation between hydrolic conductivity and matric suction in silty soils with respect to saturation. In: 17th National Soil Mechanics and Geotechnical Engineering Conference; 2018

13. Fredlund D, Xing A. Equations for the soil-water characteristic curve. Canadian Geotechnical Journal. 1994;31:521-532

14. Brooks AT, Corey RH. Hydraulic properties of porous media and their relation to drainage design. Transactions of ASAE. 1964;7:26-28

15. Van Genuchten MT. A closed-form equation for predicting the hydraulic conductivity of unsaturated soils. Soil Science Society of America Journal. 1980;44:892-898

16. Vaunat J, Romero E. An elastoplastic hydro-mechanical model for unsaturated soils. In: Tarantino A, Mancuso C, editors. Experimental Evidence and Theoretical Approaches in Unsaturated Soils. Rotterdam: Taylor & Francis; 2000. pp. 121-138. ISBN: 90 5809 186 4

17. Gens A, Sheng D, Sloan SW. A constitutive model for unsaturated soils: Thermomechanical and computational aspects. Computational Mechanics. 2004:33:453-465

18. Tamagnini R. An extended Cam-clay model for unsaturated soils with hydraulic hysteresis. Geotechnique. 2004;54:223-228

19. Sheng D. Review of fundamental principles in modelling unsaturated soil behaviour. Computers and Geotechnics. 2011;38:757-776. DOI: 10.1016/j.compgeo.2011.05.002

20. Eyupgiller M, Ulker MBC. Effect of plastic deviatoric strains on the hardening of unsaturated soils in relation to their hydromechanical behavior. In: 8th National Geotechnical Symposium; 2019

21. Ulker MBC. Constitutive modeling of cyclic seabed behavior around coastal and offshore structures: Two robust models and their predicting capabilities. In: Proc. 29th International Ocean and Polar Engineering Conference; Honolulu, HI: ISOPE; 2019

22. Baksi HG, Ülker MBC. Analysis of dynamic response of a caisson type gravity quay wall—Seabed soil system under waves. In: Proceedings of the International Offshore and Polar Engineering Conference; 2017

23. Ülker MBC, Fard MM. Instantaneous liquefaction of seabed around rubble mound breakwater under waves. In: Proceedings of the International Offshore and Polar Engineering Conference; 2016

24. Biot MA. General theory of three-dimensional consolidation. Journal of Applied Physics. 1941;12:155-164. DOI: 10.1063/1.1712886

25. Biot MA. Theory of elasticity and consolidation for a porous anisotropic solid. Journal of Applied Physics. 1955;26:182-185. DOI: 10.1063/1.1721956

26. Ulker MBC, Rahman MS. Response of saturated and nearly saturated porous media: Different formulations and their applicability. International Journal for Numerical and Analytical Methods in Geomechanics. 2009;33:633-664. DOI: 10.1002/nag.739

27. Ulker MBC, Rahman MS, Jeng DS. Wave-induced response of seabed: Various formulations and their applicability. Applied Ocean Research. 2009;31:12-24. DOI: 10.1016/j.apor.2009.03.003

28. Ulker MBC. Pore pressure, stress distributions, and instantaneous liquefaction of two-layer soil under waves. Journal of Waterway, Port, Coastal, and Ocean Engineering. 2012;138:435-450. DOI: 10.1061/(ASCE)WW.1943-5460.0000155

29. Ülker MBC. Wave-induced dynamic response of saturated multi-layer porous media: Analytical solutions and validity regions of various formulations in non-dimensional parametric space. Soil Dynamics and Earthquake Engineering. 2014;66:352-367. DOI: 10.1016/j.soildyn.2014.08.005

30. Ülker MBC. Instantaneous liquefaction in a multi-layer seabed under nonlinear progressive waves. AIP Conference Proceedings. 2012:1479-1484. DOI: 10.1063/1.4756444

31. Ulker MBC, Tatlioglu E, Lav MA. Dynamic response and liquefaction analysis of seabed-rubble mound breakwater system under waves. Applied Ocean Research. 2018;78:75-87. DOI: 10.1016/j.apor.2018.06.008

02 STABILITY ANALYSIS OF ROMAN ROCK-CUT TOMBS IN ALEXANDRIA, EGYPT: UNIAXIAL AND TRIAXIAL CREEP PERFORMANCE OF CALCARENITIC AND SANDY OOLITIC LIMESTONE FORMATIONS

1. INTRODUCTION: AN OVERVIEW

Creep is a time-dependent, irreversible ductile deformation that does not result in fracture, and it occurs gradually rather than abruptly when stress is applied, as opposed to brittle fracture. Rather, strain accumulates as a consequence of prolonged tension [1, 2, 3, 4]. Rock salt, shale, buttocks, venetian, silt, and sandstone are fragile rocks that are typically identified by this behavior. The literature has extensively discussed rock creep behavior, which is based on experimental results from laboratory or field investigations, foundational modeling, and numerical analyses. The primary objective of this chapter is to concentrate on the most suitable methodologies for establishing the creep behavior of soft/weak intact rocks by conducting laboratory experimental analysis and conducting a critical evaluation of the available rheological models to elucidate creep behavior [5, 6, 7, 8, 9]. The deformation rate is too sluggish, which is why creeping in fragile hard rocks is uncommon. Creep behavior in solid minerals is only apparent at temperatures and pressures that are typically not encountered in engineering structures. On the other hand, soft materials primarily creep at room temperature, atmospheric pressure, and the range of deviating stress that is typically encountered in engineering structures [10, 11, 12]. As we are aware, the long-term stability of the rocks and the adjacent surface is significantly impacted by creeping rocks [13, 14, 15, 16]. Porosity is the primary determinant of creep characteristics in broken rocks; however, the stress rate was the primary method used to define the creeping properties of broken rocks in the existing literature. For example, Wang [17, 18] conducted numerical simulations to examine the process of creeping damage to the road surrounding the rock under high pressure. Zhu and Ye compared the results of the rock creep test under dry and saturated conditions to examine the law of creep as it applies to water content. The properties of creeping pebbles were investigated by Zhang and Luo [19] in response to varying stress levels. Triaxial creep experiments were conducted on coal and rock by Liu et al. [20] using the step loading method. Parkin [21] employed a pressure meter to investigate the rheological properties of granular materials, while Zhang and Luo [19] conducted distinct creeping tests on marble and soft rock. Shen and Zhao [22] conducted rheological experiments on limestone to develop a model for three parameters of encroaching rock filling. Zheng and Ding proposed a creep model for rocks with nine parameters and obtained parameter indexes through experiments. A modified three-parameter rheological model for coarse-grained materials was proposed by Guo et al. [23]. Wang [24] and Liu et al. [25] summarized the rheological status of coarse-grained materials and observed that experimental investigations on granular materials were insufficient. They recommended that the effect of the scale on internal testing necessitate the investigation of the mechanism of partial deformation in coarse granular materials. A thorough examination of the mechanical behavior and engineering properties of the calcarenite rocks is necessary to comprehend the mechanisms of deterioration in the calcarenite rock structures where the Greco-Roman monuments are excavated. Numerous investigations have been conducted on rock degradation and disintegration, in addition to geological and geomorphological concerns.

Catacombs of kom El-Shoqafa.

El-Shatbi Necropolis.

Moustafa Kamil underground tombs.

Amod El-Sawari site (library and Serapeum temple).

Figure 1. Underground monuments (Catacombs) in Alexandria, (present conditions).

Sampling is permitted in a restricted number of locations with official authorization, as the areas are an open museum and popular destinations for visitors. Figure 1 illustrates the physical, short-term, and long-term mechanical properties of calcarenitic rocks in the laboratory. To achieve this, cylindrical samples with a diameter of 42–44 mm and a height of 90–100 mm were prepared using a basic drilling machine and some blocks collected from archeological sites under investigation (catacomb from Kom El-Shoqafa, El-Shatby tombs, and tombs of Mustafa Kamel). A number of samples were used for testing, and the limitations of the number of blocks were overcome by determining the topical properties of the rocks through hammer tests. Schmidt, pictorial geographic investigations and classification of the stony hill in certain outcrops and rock structures where testing was permitted.

The objective of this investigation is to provide suggestions for the reinforcement and safety of archeological subterranean structures that are subjected to both short- and long-term loading. A series of experimental experiments and sophisticated digital analyses were implemented to achieve this objective. Calcarenitic rocks and other types of fine limestone (currently under investigation) are porous rocks with intricate behavior [26, 27, 28]. It is possible to identify two primary mechanisms that can distort the properties of certain types of rock, depending on the in-situ stress conditions: (1) the prevalence of fracture, which is associated with volumetric expansion and fragile behavior and is predominant in compressive stress paths in the absence of low confined pressure, or (2) pore breakdown, which dominates high-stress conditions and produces plastic deformations and large contracting [29]. The structural heterogeneity evident in the variance of mechanical properties and deficiencies in the conclusion of experimental results is primarily attributable to the high fossil content, which is primarily attributed to the shells of necrosis and certain mollusks [13]. The establishment of the initiation and development of stress-induced fractures in EDZ is a significant concern, as there is no generally accepted theory of fragile rock strength that is founded on the examination of the process of microcrack formation and deformation. Some of the primary concerns regarding the stability of underground structures in soft rocks include the potential land disturbances caused by the drilling method and the reallocation of pressures at the site encircling the excavations [30, 31, 32, 33, 34, 35]. The stability of the drilling boundaries and the increasing permeability of host rocks to the near field can be adversely affected by each of these factors, which are related to the initiation and spread of brittle fractures and the extent of the troubled drilling area (EDZ). In the fields of structural and tectonic geology, the evolution of natural structures and tectonic features is significantly influenced by experimental rock deformation [36, 37]. A significant amount of effort has been devoted to the comprehension of the processes and mechanisms of fragile fractures. A significant portion of this attention was devoted to the quantification and measurement of fragile fracture thresholds through laboratory experiments [7, 38]. The onset of expansion, which is the reflection point of the volumetric pressure curve, is a particularly significant indicator of damaged thresholds. This is due to the fact that numerous studies have linked the threshold to the spread of unstable fracture in brittle rocks [7]. The unstable crack spreading is the moment at which the reproduction process is regulated by the speed of crack growth and the applied stress. In the event that the applied load ceases and remains stable, the fracture will persist in its propagation until it reaches failure. Martin and Chandler and Read et al., therefore, equated the long-term on-site strength of fragile rocks with the threshold of harm caused by cracking. Consequently, it is imperative to identify these processes and their corresponding mechanisms in order to accurately predict the strength of soft materials in the short and long term. The results of numerous short- and long-term laboratory experiments are presented in this research, which concentrates on these processes. In general, the irreversible destruction of molecular cohesion along the path of the fracture generated can be equated with the spread of cracks. In this regard, the rock material is "damaged" by the miniature pulverizing process. The damage can be regarded as cumulative and may be linked to a perceived deficiency in elastic rigidity and the strength of material cohesion as a result of the multiplication of the number of reproductive fractures. In this study, we emphasized several critical attributes of the geotechnical behavior of porous rocks that are structured and demonstrated that these characteristics are prevalent in a variety of natural rocks. In the past two decades, there has been a significant increase in the study of soil/rocky transition material, as a result of these concepts [39].

2. GEOLOGICAL AND TECTONIC SETTING

The Roman underground tombs in Alexandria are located on the northern edge of the "Nile Delta geomorphic province, c. 1.30 km north of the Lake Maryout and 1.42 km south of the Mediterranean Sea shoreline, as shown in Figure 2". Since Pleistocene time, within the last 1 million years, Lake Maryout has intermittently been connected to fresh Nile river flows and sea water sources and has been both at and below mean sea level. Lake Maryout and Delta had varied depositional environment, including "silt and clay deposits with some organics (lagoonal deposit)"; "sand and silt deposits (Nile River deposits); "sand deposits (beach and littoral deposits"). The basement rock unit is Miocene (6–25 million years old) and older carbonate formations that comprise the Egyptian plateau. Above the Miocene sedimentary rocks are Plio-Pleistocene age (less than 6 million years old) sediments consisting of alternating beds of shale, limestone, sandstone, silt, and calcareous sand.

Figure 2. The limestone outcrop at the catacombs of Kom El-Shoqafa.

The Plio-Pleistocene sediments form a series of ridge and trough that are approximately parallel to the Mediterranean coastline in the vicinity of the catacomb site. Most of the city of Alexandria rests on one of these topographic ridges while behind the ridge, Lake Maryout is in a trough. The near surface limestone deposits, which are commonly encountered in the Alexandrian ridge, are cemented marine sand.

3. STATE OF PRESERVATION

The catacombs of Alexandria show some clear indications of yield and partial collapse in several locations, as defined in the honeycomb weathering, the contour scaling and spalling of the stone surface, the disintegration of building materials, and the wet surfaces of rocky meals especially for semi-protected parts of the excavation; also, we observe salt flowering and yellow staining of yellow iron in many wall parts.

Structural damage is obvious like the wall cracking, the thinning out of rock pillars, disintegration and degradation of the walls surfaces, the partial collapse of some parts of the roofs and walls, and the peeling of rocks, especially in the roof of narrow corridors found in the deepest parts and mass waste from the ceiling and walls.

In conclusion, the current state of conservation of the great catacombs at Kom El-Shoqafa, the best-known and most famous testimony of the culture of the funerary architecture of Alexandria, is now at its most deteriorating.

Most structural damage is caused by one or a combination of the following factors:

- The gradual weakening of rock materials due to the intrinsic sensitivity of weathering factors, especially the effect of weathering with groundwater and salt
- Earthquake and other man-made dynamic loading
- Permanent deformation of the rock mass
- Natural wear and tear of materials
- History of construction in the area

4. MINERALOGICAL AND PETROGRAPHICAL STUDIES

The effort behind thin-section analysis was to provide insight into the closed grains (calcite/sand) and/or theories of overgrowth after precipitation of the large angle of internal friction. Due to the fragile nature of the rocks and plaster layers being excavated, it was necessary to be very careful to make thin sections, which were studied using independent polarized light, electron microscopy (SEM), and stereoscopic observation.

A light-transmitted polarized plane, scanning electron microscopy, and stereoscopic observations were used to determine the interlocking textures and connections between grains and crystals. These contacts rely on differences in solubility due to impurities and differences in bending radius, which lead to the penetration of smaller grains in large grains. In addition, thin section microscopy was used to help explain the large friction angles associated with the material, limestone/rock.

Miniature petrographic description of stones/rocks for engineering purposes includes the identification of all parameters that cannot be obtained from a comprehensive endoscopic examination of rock samples, such as mineral content, grain size and texture, which have an impact on the mechanical behavior of the rock or rock mass. To ensure proper classification, the first step should be to check the metal composition and rock texture; see Table 1. Mineralogy summarizes the three types of soft limestone under investigation. Additional investigations should include analysis of the texture and minerals in the case of highly contrasting rocks, determining the degree of change or weathering, grain size, partial fracture, and porosity.

In sandstone, limestone, and calcarenite samples intact, it is possible to determine with the naked eye an alternative sequence of white and pink bands with a thickness of about 1 mm (bedding plane). Optical microscopy and counting points were performed on thin sections of rock samples. The air-dried samples were inoculated with Canada balsam, and the thin sections were then cut perpendicular to the bedding planes. A thin section is observed under parallel light and polarizing light. The following is a detailed analysis of the rock samples collected from the three archeological sites under investigation, rock samples from six collections of El-Shatby with code Nr (SH), five rock samples collected from the tombs of Mustafa Kamel 1 and No. 2 with code Nr (M), and four samples Rock collected from Catacomb of Kom El-Shoqafa code Nr (COM).

Table 1. Mineralogy of the three soft limestone types under investigation.

Rock type	Calcite $CaCO_3$ %	Quartz SiO_2 %	Gypsum $CaSO_4.2H_2O$ %	Halite NaCl %	Other %
Sandy oolitic limestone (Kom El-Shoqafa) (COM)	47–65	31–23	10–5	12–9	2
Intact Calcarenite (Mustafa Kamel Necropolis) (M)	52–72	28–18	8–3	12–6	3–5
Oolitic intraclastic limestone (El-Shatby Necropolis) (SH)	53–80	25–20	11–7	11–7	3

4.1 Catacombs of Kom El-Shoqafa

In the internal structure, we can observe the dominant components, which are the cells of the fibers of the stomach, grass, algae, and mother of pearl, mostly with a test wall of microscopic microspheres, while the tests are filled internally with microtomes and microbes (Figure 3). Surrounded monocrystalline quartz granules of varying sizes and perimeter of iron oxides have been detected representing the previous presence of K-feldspar grains. Rock and granular materials make up this fossil sand limestone, or cement sand.

(Calcarenite size) 15% of customizations are medium-sized numulite tests filled with prickly calcite. 15% of foraminifers tests with a neomorphic microspar test wall and test chambers are full of neomorphic microspar. 20% of medium size bryoza and algae tests 0.25% small size, monocrystalline, crispy extinction, quartz granules subrounded. 25% medium to small size structure less ooides. Customizations are solidified by isopachous microspar. Porosity is a fit of 20% of the area of the thin-section field, which is reduced by microscopy. Oxidation is observed as red color spots.

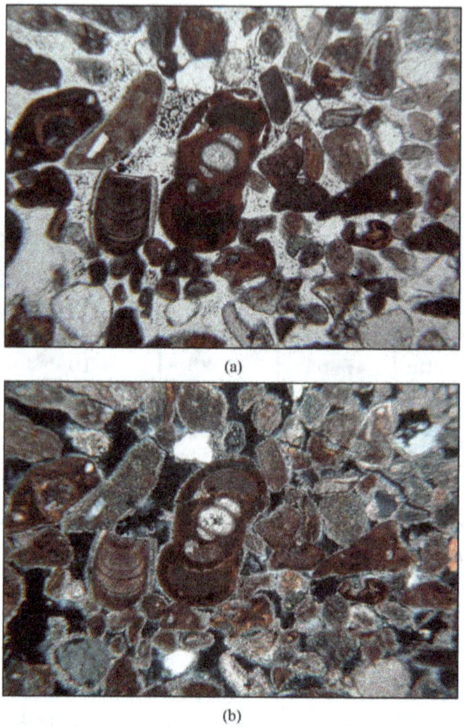

Figure 3. Photomicrograph of fossiliferous sandy oolitic limestone, (a) under parallel polarized light, (b) under cross polarized light (XPL), showing bioclasts of gastropods, foraminifera, algae, and shell debris; most of them are with test wall of neomorphic microspar, filled with micrite and microspar, cracks between and through the minerals are obvious. Catacombs of Kom El-Shoqafa.

4.2 Mustafa Kamel Necropolis

The rock texture in these tombs consists of two textures, namely packed stone and stone. These two types of texture show different proportions and sizes of quartz granules, and different biological plates, especially foraminifer tests. Most Ooides lost their internal structure. Few of them retain their concentric structure. Consolidation of the components of this limestone is represented by isopachous microspar (Figure 4).

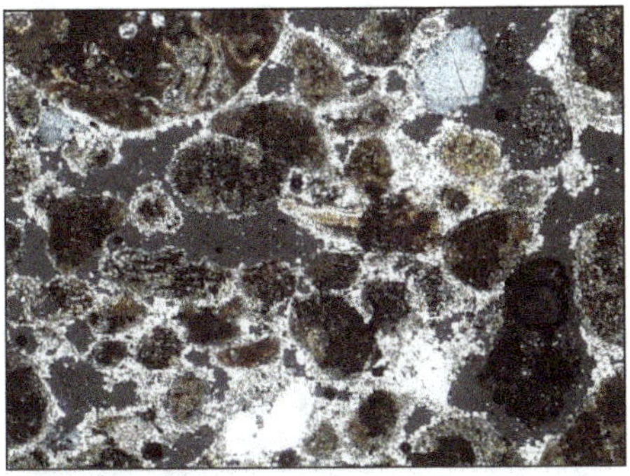

(a)

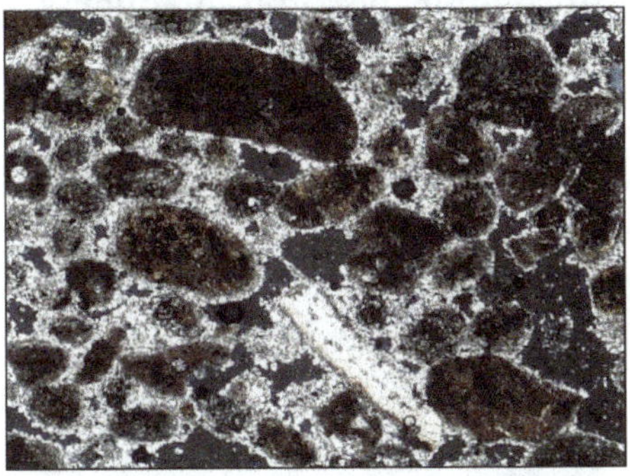

(b)

Figure 4. (a, b) Photomicrograph of intact calcarenite under cross polarized light (XPL) showing wackestone (pele-oo-sparite) texture with drusy sparite, Mustafa Kamel Necropolis (Weathered sample, heterogeneous pore system).

(Calc rud –arenite size) 58% of the assignments are medium in size, thin and micro pigment and less pollutant internal structure. 10% micritic oval. 30% large to small angular size to subrounded, crispy extinction, monocrystalline quartz. 2% plajioclase and microcline crystals. Porosity reached 20% of the area of the thin-section field. The pores are filled with neomorphic microspar. Allochems are surrounded with isopachous microspar.

Calcarenite is a bio-soft rock originating from marine sediments, which occurred during the overflow and decline of the region in the Ice Age. The calcarenite consists of almost pure calcium carbonate and is applied directly to the limestone rock of the Cretaceous.

4.3 El-Shatby cemetery

Changes in internal structure and metals were analyzed and the most distinctive textures documented on the images. In the internal structure, we can observe the porosity increase of various sizes. In some places, we can find cracks on metal contacts or even inside metals. Generally, significant changes are shown in the cement material; see Figure 5. Limestone in this site can be classified into two types of fabric, namely, fossiliferous oolitic intraclasic limestone. These two types of texture are in different proportions of quartz granules, biological panels, ooides, and peloids.

(a)

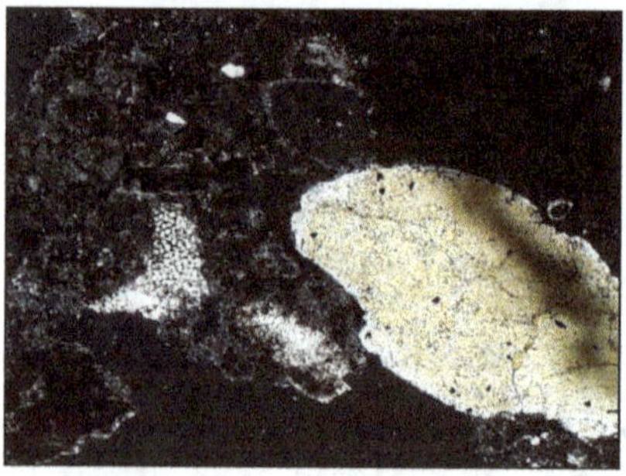

(b)

Figure 5. (a, b) Photomicrograph of fossiliferous oolitic intraclastic limestone thin section under cross-polarized light (XPL) showing subrounded monocrystalline quartz grains (QTZ) and porous region, El-Shatby Necropolis.

(Callus arinite rod size) 80% of the customizations are medium-sized structure less ooides. 10% large to medium-sized monocrystalline unite extinction, quartz granules subrounded. 5% large polycrystalline, crispy extinction, quartz granules subrounded. Five% of algae and foraminifera are tested with a micritic wall and are filled internally with microscopic grains. Porosity is greatly reduced due to their filling with depressed dwarfs. The new form is observed to worsen from micrite to microspar. The evaluation of thin sections allows the analysis of pore structure and enables the assessment of pore size and distribution in relation to the distribution and formation of the minerals involved.

4.4 Comparison between the sound and weathered rock layers

Samples can be clearly distinguished from the alveolar portions - the amortized and non-woven parts using thin, unpainted limestone sections that feature a relatively homogeneous pore structure. In contrast to unpainted areas, alveolar flats have a heterogeneous pore structure, for example pores often contain ferric oxides and hydroxides indicating a lower total pore size and higher content of small spots.

Data from microscopic polarization and electron microscopy experiments show that oxygen clarity of NaCl crystals is strongly influenced by the rate and volume of moisture changes, and how they shrink with changes in crystal size.

5. CREEP TESTS (MATERIALS AND EXPERIMENTAL PROGRAM)

Creep is an irreversible ductile deformation in time under constant stress. Creep strain seldom can be recovered fully when the loads are removed, thus it is largely “plastic deformation.” It is a progressive phenomenon initiated at a certain time after excavation at a certain location around the profile and spreading in time into the rock mass. For the long-duration design life of underground structures, the long-term stability of the tunnel must receive major consideration. For this reason, time-dependent deformation behavior of the surrounding rock must be well understood. Neglecting creep effects during deep excavation may lead to incorrect evaluation of deformation and thus may impact on the criteria for selection of proper design.

Understanding the mechanisms of rock breakdown that have been excavated within ancient monuments requires a thorough study of the mechanical behavior of these rocks, and the importance of the physical and mechanical properties of these rocks to understand the phenomena of instability.

The results of the geotechnical characterization of these rocks will be used in numerical modeling and design of reinforcement measures. For this purpose, a new laboratory testing program will be launched.

Rocks, sample preparation, experimental setup used and the procedure are briefly described below.

5.1 Types of creep

An idealized creep curve for rock at constant stress consists of three stages: instantaneous elastic strain followed by primary creep with decreasing creep rate, then steady-state creep with constant creep rate, and finally tertiary creep with increasing creep rate leading to failure. Most of the work on time-dependent strain has been conducted on primary and secondary creep phases only and the tertiary phase has not been investigated in appreciable detail.

5.2 Laboratory test specimens

In this study, the size of the comprehensive laboratory testing program using cylindrical samples with 42-44 mm diameter and height (91–103 mm). Although these rocks do not show distinct layers, the nuclei were extracted from the blocks and their masses in the vertical direction, which was expected to represent the physical properties of these units perpendicular to the layers. However, some samples were also extracted in a vertical direction on the mattress. Some specimens were broken and/or small cracks or cracks appeared on their surfaces. However, in order to achieve reliable assessments, the number of samples was increased as many as possible. Laboratory tests were performed in accordance with the testing procedures proposed ISRM and recommended by ASTM at the Engineering Geology Laboratory, Department of Civil Engineering, University of Aristotle Thessaloniki, Greece.

5.3 Laboratory tests

Laboratory studies (experimental examination) were performed on surface rock samples and prepared surfaces. The basic mechanical testing of the laboratory includes the behavior of deformation to failure under uniaxial and triaxial compression and we offer a complete creeping rock characterization conducted during the past 2 years from a series of isotropic and isotropic compression tests conducted in the inventory of various stresses, viscosity behavior was determined by following a procedure, the multi-step download, which emphasizes the transit creep side.

5.4 Very slow uniaxial creep tests on calcarenitic and sandy oolitic rock specimens under investigation

Creep in hard brittle rocks is rare as deformation rate is extremely slow. Hard rock shows creep behavior appreciably only at elevated temperatures and pressures generally not encountered in engineering structures. Soft rocks on the other hand creep mostly at the room temperature, atmospheric pressure, and deviatoric stress range normally encountered in engineering structures.

5.5 Analysis of creep behavior of soft rocks in tunneling

Regarding viscous plasticity, despite much work done on high porous rocks, only over the past years, there has been growing concern about the long-term behavior of deep underground structures in general. The rock mass tests large strain rates of viscosity and plastic. However, after a few years, the stress rates become smaller and reach a fairly stable condition characterized by very small stress rates.

It is known that most rocks have time-dependent behavior, and the viscous and plastic modeling of rocks and soils is of great importance both in petroleum engineering and underground engineering, for example when assessing deformations at the walls of deep fossil sections or considering pressure problems.

Moreover, when smaller time periods are considered, the stress distribution around a cave or exposures is such that the divergent pressure decreases rapidly with respect to the distance to the cave. Very small stress rates are tested at large distances within the rock mass and should be evaluated when predicting the behavior of the cave or photo gallery [40].

The limited available literature may be rooted in the particular problems raised by the long-term creep test, in the short term, as described below.

When the creep rate is 10 = 10–12 s–1, a 12-day test results in a strain of = 10–6. The coefficient of thermal expansion of rocks is in order α = 1–4 × 10–5 C–1, that is, the "noise" (i.e., elastic thermal deformation sample) due to small temperature changes will be greater, in most cases, than the signal to be measured (e.g., the average sample deformation arose from proper creep). The same can be said for moisture variations, which have a significant impact on many rocks.

- Slow creep rates are obtained when small mechanical loads are applied. Most of the crawl test devices are designed to work in a DVR pressure range of 5–20 MPa. Stress control is usually weak when the applied pressure is less than 1 MPa.

The creep rate is calculated by comparing strains 1, measured in two different times, τ1 and τ2, or . = (2 - 1) / (τ2 - τ1). When the compression rate is in the range = 10–12 s–1, it can be reasonably evaluated on a daily basis (t2 - t1 = 105 s, 2 - 1 = 10–7) only if 1 and 2 can be measured with an accuracy of not less than 10.8, or one-tenth of the expected difference between the two successive measured breeds [40].

Tightening of fragile rocks results in distributed damage long before the rocks fail unstable. The damage is usually manifested in small fractures and expansive microcracks [41, 42, 43]. These small fractions are usually smaller than the grain size and are often distributed almost uniformly before they are locally cracked. There are no uniform distributions of small fractions associated with the nucleus of error and growth.

Partial damage was used to explain the reduction of seismic wave velocity, earthquake variation, reduction of elasticity and strength units, and rock failure mechanics. In addition, stress damage can facilitate time-based creep-driven by stress erosion and subcritical crack growth. This creep strongly affects long-term strength and failure stability. For example, granite samples that are exposed to 1 month of non-axial static pressure under a pressure of approximately 0.65 may fail—or "delayed fractures" may develop days to years after removal of applicable loads.

The creep test shows how strain builds up over time under constant pressure. The rock usually deforms quickly and then begins to deform more slowly after the yield fatigue, which is called the initial creep. After the initial creep (I), the deformation continues at a constant rate in the linear part of the curve, which is secondary creep (II). Finally, the deformation rate increases rapidly until the rock fails to "fracture" in the high creep (III), if stress is removed but the strain remains permanent.

Three stages of creep behavior can be identified: in the first stage, they are classified as initial creep, and strain occurs at a decreasing rate. In some cases, the primary creep curve approaches a constant rate of strain called secondary creep. In high-stress specimens, secondary creep may turn up in higher creep, which is characterized by an increased strain rate until creep failure occurs suddenly. In the last two stages, the thin vertical cracking begins, accompanied by hardening, and only near failure, large cracks spread rapidly and lead to a sudden collapse. Long-term tests performed on a secondary creep sample revealed even appearance at 40% of estimated strength. The purpose of this research is to make recommendations on the promotion and safety of long-term underground historical structures under load. For this purpose, there is a set of experimental tests and advanced numerical analyses.

6. DESCRIPTION AND DISCUSSION OF THE EXPERIMENTAL PROGRAM

6.1 Describe full creep tests

The research demonstrates an integrated empirical approach aimed at assessing safety and strengthening historic underground structures under high pressure.

The purpose of these tests is to obtain data, first, to determine the amount of sticky parameters that govern the long-term behavior of these structures, and secondly, to validate numerical models.

6.2 Testing device

Long-range uniaxial creep tests were performed on standard cylindrical rock samples collected from the three archeological sites under investigation (diameter D = 4.2–4.4 mm, height H = 90–103 mm); samples were prepared for testing according to ASTM standards with length-to-diameter ratios approximately 2.25, all samples have highly polished end surfaces to minimize final effects. The sample was set between two solid steel plates, with a steel cover between the sample and the two plates. During each test, two high-precision displacement sensors at two vertical levels at a 90° angle allowed both the relative rotation of the two pages and the measurement of the average relative displacement.

6.3 Sensors

Applied loads and the resulting strain were recorded using an automatic data acquisition system, sampling at a rate between 1 and 3 readings per second, thereby overcoming any deficiencies in data resolution.

6.4 Loading

The approved test procedure consisted of loading samples at a constant rate of about 1.35 MPa up to 1.75 MPa for samples from Catacomb in Kom El-Shoqafa, 1.55 MPa up to 2.17 MPa for samples from Mustafa Kamel Necropolis, and 2.6 MPa up to 3.44 MPa for samples from El-Shatby cemetery. In order to keep the applied pressure as stable as possible, dead weights were used and steel cylinders were placed on the upper steel plate on the upper face of the cylindrical sample. The applied stress is calculated by dividing the weight of the steel cylinders placed on the top plate by the initial cross-sectional area of the sample.

6.5 Temperature and hygrometry

The temperature changes during a long-term creep test must be as a small as possible and must be measured precisely enough to allow correction of the raw strain data for thermoelastic strains; in our study, all the periods of test were in the room temperature between 24 and 26° in the laboratory by controlling the air condition.

7. TEST RESULTS

Uniaxial creep tests were performed on three rock samples from each site. Rock samples are loaded through fixed uniaxial compression at 1, 35, 1, and 75 MPa (one stress per sample) for Catacomb of Kom El-Shoqafa rock samples collected, at 2.6 and 3.44 MPa for rock samples collected from El-Shatby archeological site, and at 1.55 and 2.17 MPa for rock samples collected from Mustafa Kamel Necropolis.

The experimental procedure follows ASTM standards (ASTM D4405 and D4341). The compression machine is used to apply continuous axial load to the samples. Digital scales are installed at 0.001 millimeters to measure the axial displacement of the samples, see Figures 6–9. Samples are loaded continuously for 1 to 2 years until the samples fail without any acceleration, depending on the displacement results. During testing, axial distortion, time, and failure modes are recorded. The readings are repeated every minute at the beginning of the test, and gradually decrease to twice a day after the first few days of testing. This also depends on the deformation rate of each sample. The results are presented by strain time curves. Axial stress and axial pressure values are calculated by:

Figure 6. Rock creep testing devices. Samples are 90–105 mm high, 42–44 mm2 diameters. Two displacement sensors were used during each test.

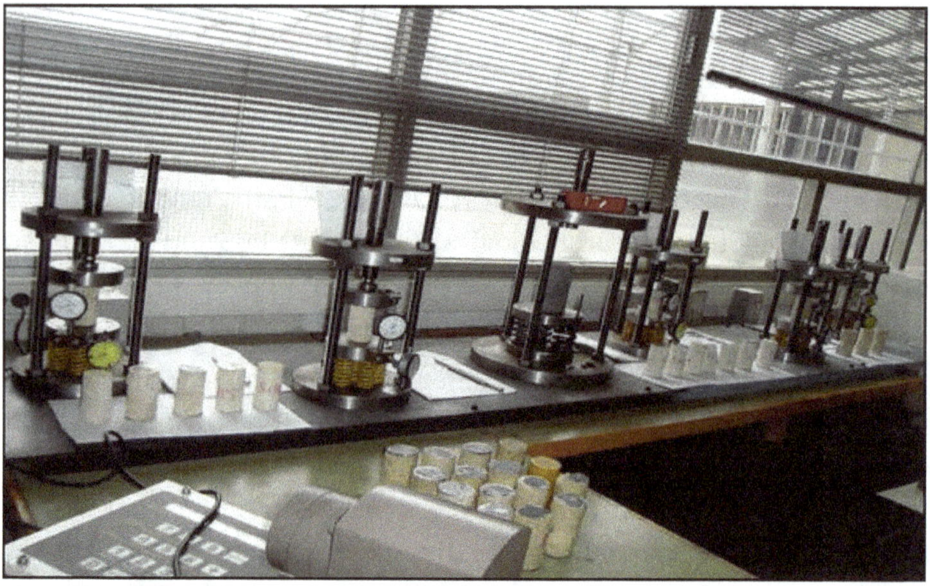

Figure 7. The collected intact sandy oolitic limestone specimens from l-Shatby Necropolis site under creep testing devices.

Figure 8. The collected sandy oolitic limestone specimens from the catacombs of Kom El-Shoqafa site under creep testing devices.

Figure 9. The collected intact calcarenitic rock specimens from Mustafa Kamel Necropolis site under creep testing devices.

$$\sigma_{axial} = Pa/A, \tag{1}$$

$$\varepsilon_{axial} = \Delta L/L, \tag{2}$$

where axial is the axial pressure, Pa is applied axial load, A is the normal cross-section area of the direction of the load, axial is the geometric axial strain, ΔL is the axial deformation, and L is the original length.

Table 2 summarizes the results of a uniaxial creep test. The axial stress time curves are shown in Figures 10–16, and the curves represent instantaneous, transient, and triple creeps of rock samples under a fixed axial load. Samples are loaded quickly and then the axial strains increase. The immediate breeds range from 0.07 to 3.5.

Table 2. Uniaxial creep test, testing program.

Specimen No.	Testing period	Time (days) 1 9 100 135 178 375 667 786 813
Catacomb of Kom El-Shoqafa test _1 (sandy oolitic limestone)	From **5/5/2016** to **1/6/2017**	**1= 1.35 MPa**
Catacomb of Kom El-Shoqafa test _2	From **1/9/2016** to **2/7/2018**	**1= 1.75 MPa**
El-Shatby Necropolis. test _1 (oolitic intraclastic limestone)	From **12/4/2016** to **4/7/2018**	**1= 2.60 MPa**
El-Shatby Necropolis. test _2 (oolitic intraclastic limestone)	From **5/5/2016** to **3/7/2018**	**1= 3.44 MPa**
Mustafa Kamel Necropolis .test _1 (intact Calcarenite)	From **3/4/2016** to **22/3/2017**	**1= 1.55 MPa** **1= 1.86 MPa**
Mustafa Kamel Necropolis. test _2 (intact Calcarenite)	From **11/4/2016** to **28/3/2017**	**1= 2 MPa**
Mustafa Kamel Necropolis. test _3 (intact Calcarenite)	From **6/4/2016** to **7/4/2016**	**1= 2.50 MPa**

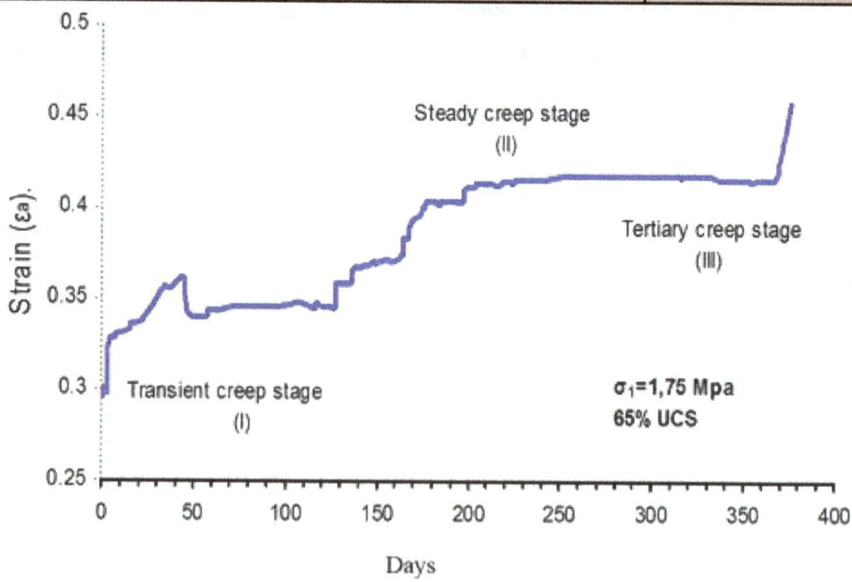

Figure 10. Strain versus time curve during the catacomb of Kom El-Shoqafa, uniaxial creep test no. 1.

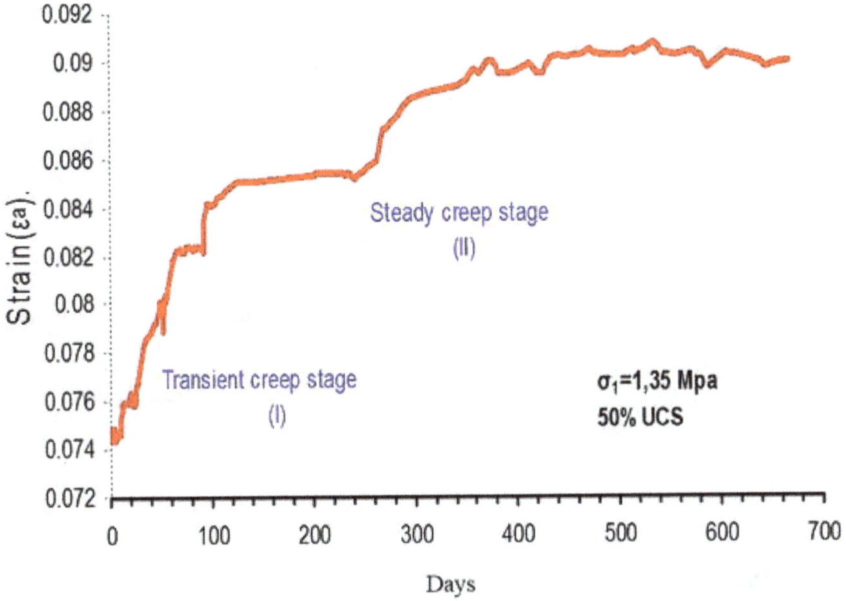

Figure 11. Strain versus time curve during the catacomb of Kom El-Shoqafa, uniaxial creep test no. 2.

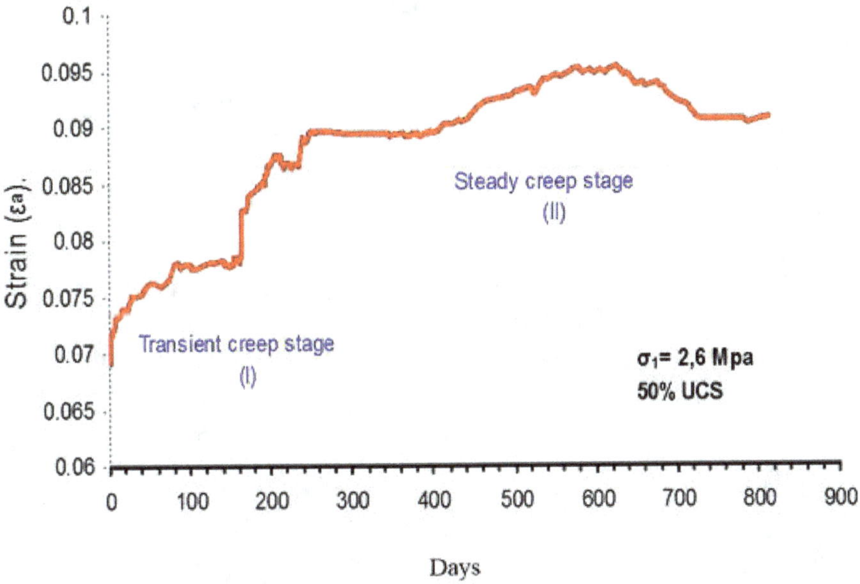

Figure 12. Strain versus time during El-Shatby Necropolis, uniaxial creep test no. 1.

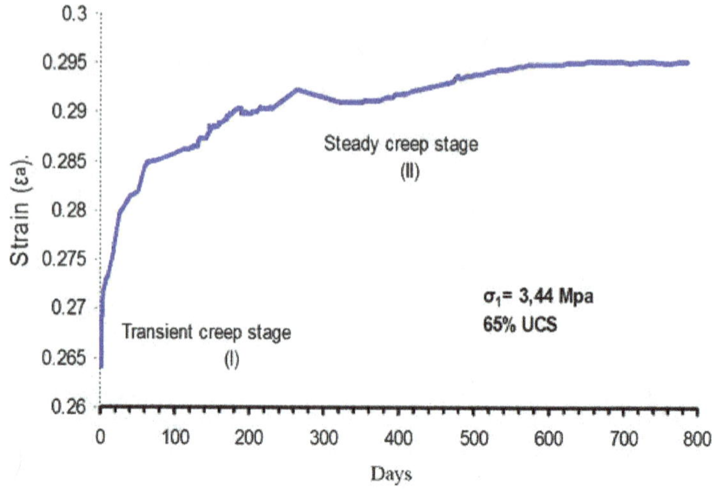

Figure 13. Strain versus time curve during El-Shatby Necropolis, uniaxial creep test no. 2.

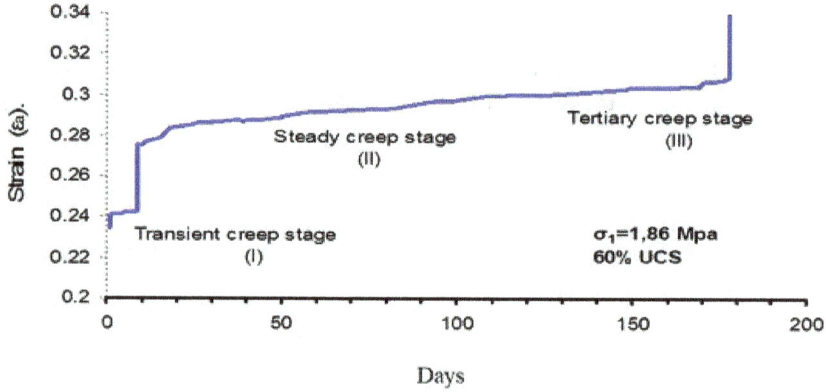

Figure 14. Strain versus time curve during Mustafa Kamel Necropolis, uniaxial creep test no. 1.

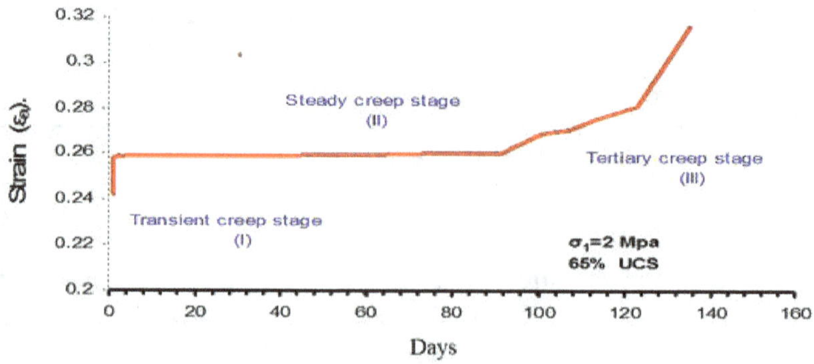

Figure 15. Strain versus time curve during Mustafa Kamel Necropolis, uniaxial creep test no. 2.

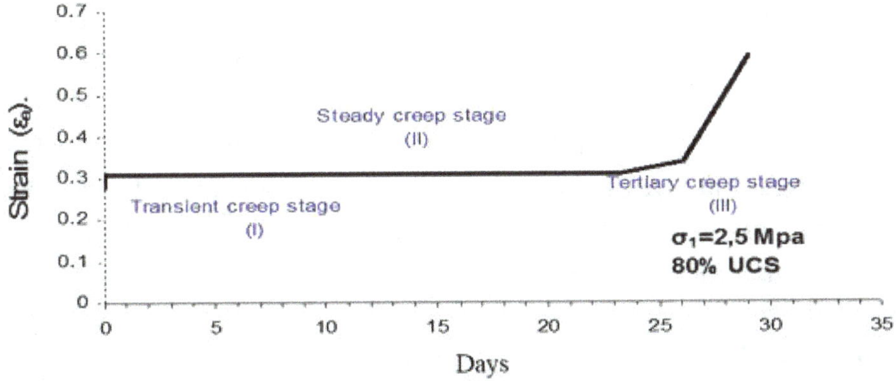

Figure 16. Strain versus time curve during Mustafa Kamel Necropolis, uniaxial creep test no. 3.

Most samples, under constant axial pressure, show a complete creep stage: transient, steady, and triple creep stages.

Increasing the value of the instantaneous creep strain with hard axial stress gives strain time curves of rock samples tested under constant high and low axial pressures. Axial stress also increases crawling strains. In the transit crawl stage, the stress rate increases with the applied stressors. In most cases, the stress rate under high axial pressure is greater than the low axial pressure rate. The effect of embedding in the sample may make the compression rate under low pressure higher than the pressure under high pressure.

7.1 Catacomb of Kom El-Shoqafa, Test No. 1

On May 5, 2016 (Day 1), Catacomb of Kom El-Shoqafa no_1 began testing on a sample of sandy limestone, loading it to a vertical stress of 1 = 1.75 MPa, 65% of the coaxial compression strength of the rock material (peak sample strength). Figure 10 displays the strain curve versus time; this curve averages the data provided by two displacement sensors. Strains do not correct for elastic thermal differences. In this test, the crawl was faster than the Catacomb of Kom El-Shoqafa site. Test no_2: From day 130 to day 200 after the start of each test, the cumulative strain was 4.5 microns for Catacomb of Kom El-Shoqafa. 1 and 2.8 microns for Catacomb of Kom El-Shoqafa test site 2. This difference is fully in line with what is known in previous tests conducted at greater pressures on these samples. When the stress rate in the transient pressure zone is increased, followed by a similar decrease, it can be observed from day 44 to day 130, immediately followed by a steady slope (steady state crawl) up to 205 days. Finally, a more stable condition followed with a smaller stress rate until the sudden sample failure on day 368. Stress rate developments were more progressive in this case. There is no specific explanation for these changes in compression rate. At the end of the test, the observed pressure rate is = 2.30 × 10−8 s−1, the sample was suddenly broken after the 368 day (end of the test) on June 1, 2017, while the sample showed a higher creep phase.

7.2 Catacomb of Kom El-Shoqafa, Test No. 2

On September 1, 2016, (Day 1) after the start of the previous tests, an identical creep device on the same table was assigned to the catacomb of Kom El-Shoqafa Test no_2, which began on another sample significantly purer than the previous, loaded on a vertical stress 1 = 1.35 Mpa, 50% of the axial compression strength of the rock material (peak strength), the applied stress until the end of the test was not adjusted without sample failure on July 2, 2018, during a steady slope or steady state and a creep with a small strain rate was observed. Figure 11 displays a curve versus time. This curve averages the data provided by two displacement sensors. The compression rate (.) is calculated every 5 days; it is calculated for 10 days. Strains are corrected for temperature variation. Initially, the strain experienced a long initial transient period until the first few days characterized by a slow decline in rate, with the average stress rate stabilizing to = 5.85 × 10−10 s−1 (positive sample contractions), with long-term amplitude fluctuations ++20%; this is probably associated with moisture fluctuations. This phase was followed by a long steady slope or steady-state creep to the end of the test while the observed compression rate was . = 3.21 × 10−9 s−1, while it was 1.50 × 10−9 s−1 was at the beginning of the test.

7.3 El-Shatby Necropolis, Test No. 1

On April 12, 2016, (Day 1) testing of Shatby Tombs No. 1 began on a sample of sound rock-limestone that was loaded to = 1 = 2.60 MPa, 50%

of the axial compression strength of the rock material (peak strength) is not Stress adjustment until the end of the test on July 4, 2018. Figure 12 shows the stress curve versus time, where the elastic strain is followed by a long transient creep characterized by a slow rate of decline, followed by a slope or creep constant in a steady state with a small stress rate until end of the test without sample failure; this curve averages the data provided by two displacement sensors. The compression rate (.) is calculated every 1 h at the beginning of the test, and after the first few days it is calculated every day. Strains are corrected for temperature variation. The strain experienced a long initial transient period, where the average stress rate stabilized on = 1.3 × 10−9 s−1 (positive sample contractions.), with long-term amplitude fluctuations ++15%; this is probably associated with moisture fluctuations.

Transient reverse crawl was observed on day 214 to day 244, sometimes referred to as "hypotension." During this test, this reverse crawl lasted much longer (20 days) than is currently observed in tests with greater stress. The stress rate stabilized one way or another after day 260, but at the end of the test, the observed pressure rate was . = 1.62 × 10−9 s−1.

7.4 El-Shatby Necropolis, Test No. 2

On May 5, 2016 (after the start of the previous tests), an identical crawl device was assigned to the same table, and El-Shatby test of Q2 was started on a cylindrical sample with geometric dimensions similar to that used in El-Shatby test of cemetery no_1, loaded on a vertical stress of 1 = 3.44 MPa, 65% of the coaxial compression strength of the rock material (Figure 13). A long transient period can be observed followed by a constant inclination or a steady-state crawl until the last day of recording. In this test, the crawl was faster than at El-Shatby Cemetery, test number 1: from day 70 to 270 after the beginning of each test, the cumulative strain was 2.5 m for El-Shatby Cemetery, test number 2 and 1.8 microns for El-Shatby Cemetery site, test number 1. This difference corresponds exactly to what is known from previous tests conducted at greater pressures on these samples. An increase in the stress rate can be observed, followed by an equivalent decrease, at day 260 and at around day 324, and stress rate developments were more progressive in this case. There is no specific explanation for these changes in compression rate. At the end of the test on July 3, 2018, the observed pressure rate was . = 3.41 × 10–10 s–1.

7.5 Necropolis of Mustafa Kamel, Test No. 1

On April 3, 2016, (Day 1) Mustafa Kamel's # 1 test began on an intact sample initially loaded at 1.55 MPa but no creep was observed until 9 days after the test began. Perhaps the pregnancy is too small to produce any detectable strain. Thereafter, the applied pressure was adjusted once, and was constructed up to 1 = 1.86 MPa (+10%) after the 9th day 60% of the uniaxial compression strength of the rock material. The numbers in parentheses indicate that the compression value is adjusted. Figure 14 shows the pressure curve versus time; this curve averages the data provided by two displacement sensors. The compression rate (.) is calculated every 5 days; it is calculated for 10 days. Strains are corrected for temperature variation. The strain experienced a long initial transient period characterized by a low slow rate followed by a steady slope or a steady-state creep with a small stress rate, at which time the average stress rate stabilized to = 1.62 × 10–9 s–1 (positive sample contractions.), with long-term capacity fluctuations of + _20%; this is probably associated with moisture fluctuations.

The transient inverse creep has not been observed, and is sometimes referred to as "stress drop." Strain rate more-or-less stabilized after day 160, and strain rate . = 4.86 × 10–9 s–1 and the sample has been broken suddenly after 178 days (the end of the test 22/3/2017); the specimen showed the complete three phases of creep end with the tertiary or acceleration creep stage.

7.6 Necropolis of Mustafa Kamel, Test No. 2

On April 11, 2016, (after the start of the previous tests) an identical crawl device was set on the same table, and Mustafa KAM # 2 test started on another sample that is significantly purer than the previous, loaded on the stress of 1 = 2 MPa, 65% of the uniaxial compression force for rocky materials, and applied pressure was not modified until the end of the test on 28/3/2017 (Figure 15). It displays the strain curve versus time; this curve averages the data provided by two displacement sensors. In this test, the creep was faster than the site of Mustafa Kamel's tombs, test number 1: from day 11 to 91 after the start of each test, the cumulative strain was 3.7 microns for the Mustafa Kamel test site Necropolis. 2 and 3.5 microns of the graves of Mustafa Kamel site No. 1. This difference corresponds exactly to what is known from previous tests conducted at greater pressures on these samples. An increase in stress rate was not observed in this test, followed by an equivalent decrease, while a long transient strain was encountered and a slow decline in rates was followed by a creeping phase in a steady state with a very small stress rate until day 91, after acceleration or the third stage of creep began. The 135th day in a large stress rate . = 1.11 × 10−9 s−1.

7.7 Necropolis of Mustafa Kamel, Test No. 3

On April 6, 2016, an identical creep device was set on the same table, and Mustafa KAM # 3 test was started on another heavily loaded sample on a stress of 1 = 2.5 MPa, 80% of the uniaxial pressure force of the material rock (peak strength). Strains do not correct for elastic thermal differences. Figure 16 displays a curve versus time. This curve averages the data provided by two displacement sensors. In this test, the crawl was faster than the site of Mustafa Kamel's tombs, test number 1 and test number: from the first day after the start of the test, the cumulative strain was 7 microns for the site of Mustafa Kamel Necropolis, test number 3. This difference corresponds exactly to what is known from the tests previously conducted at smaller pressures on these samples. The sample fractured 26 h after the start of the test, the crawl begins with a short elastic strain followed by a short transient strain followed by a steady-state crawl with a very small stress rate up to 23 h after the start of acceleration or triple crawl resulting in a sudden failure of the sample with a high stress rate after 26 h exactly. At the end of the test, the observed pressure rate was . = 0.30 per second.

7.8 Comparison with tests performed under larger stresses

Qualitatively, the behavior of soft rocks under small pressure (0.1 = 0.1–3 MPa) exhibits the same general features as observed under large pressures (e.g., = 5–20 MPa). The rapid accumulation of stress leads to a transient creep characterized by a slow rate of decline. The creep rate then becomes almost constant (a steady state is reached) or, more precisely, its average value remains constant, but the rate faces long-term fluctuations that may be affected by slow changes in moisture measurement. Reducing the load ("low pressure") creates an inverse crawl, which lasts much longer during tests under greater stress.

Norton-Hoff's constitutive equation is often proposed to describe stable state creep.

$$\varepsilon^{\cdot} = A^{*} \ \exp.(-Q/RT)\,\sigma^{n} \tag{3}$$

where is the applied deviatoric stress; T is the absolute temperature; and A*, n, and Q/R are constants. For Etrez salt, Pouya suggests the following parameter values:

$$A^{*} = 0.64\ \mathrm{MPa}^{-n}\ \mathrm{yr.}^{-1}, Q/R = 4100\ \mathrm{K}, \text{and } n = 3.1. \tag{4}$$

Berest et al. [40] found that if the Norton-Hoff Law of Conditions was derived in Creep Test 1 (=0.108 Pa, T=286.5 K), the calculated compression rate (. = 10-17 s-1) is smaller. Start by from the observed compression rate (. = 1.4 × 10-12 s-1). The observed pressure rates, even if they are too small, are much larger than expected. Spears et al. suggest that the pressure solution (rather than infiltration and slip, the mechanism that controls high stresses) is the most effective mechanism for crawling at very small pressures; the exponent of stress in this context would be n = 1 instead of n = 3–5, which is observed during standard tests. If this proposal is adopted, the Creep law should be modified when considering small pressures, with significant consequences in predicting the cave or gallery convergence rate.

Many lessons were learned during the test under these unusually low pressures. This first series of tests opened the way for further research on the behavior of rocks under very small pressures, long-term single-axis crawl tests were performed for geological and engineering applications on rock samples (for 850 days), and the applied loads were as small as 1.35 MPa. Slow stress rates such as 1.11 × 10-10 s-1 were observed in some cases. These small loads and pressure rates pose several specific problems: potential drift of sensors during long 2-year tests, interference with small changes in room temperature and moisture measurement, and effects related to irregular load distribution applied to sample surfaces. These difficulties have been recognized and at least partially addressed. The qualitative results are in good agreement with what is known as the behavior of soft rocks under greater pressure; however, the observed pressure rates, even if they are extremely small, were much greater than expected.

The initiation, accumulation, and growth of cracks caused by stress in rocks are generally referred to as rock damage. Referring to the pressure caused by the crack is the load at which the sample will eventually fail, under prolonged loading, which they propose correspond to about 70–80% of the peak strength of the sample. It is also believed that the damage to the crack damage or the crack damage threshold point corresponds to the point at which the stress reflection or sample expansion begins. Corresponding to the volumetric stress gradient is approximately 70% of the estimated unrestricted compressive strength of the rock.

These stresses are well above the stress threshold for damage. It has been suggested that sample composition for unrestricted compression force tests reduces the spread of cracks. Many researchers suggest that the strain of the ring is generated between a stretch crack and the outer surface of a cylindrical sample. This breed may generate a confinement collar that limits the growth of continuous cracks.

Preliminary test results suggest that an alternative mechanism may affect the spread of unstable cracks. Under pure uniaxial loading conditions, a split can be expected parallel to the maximum pressure direction. The failure may ultimately be at the microscopic level due to the curvature of the rock slabs resulting from tensile fractures directed toward the maximum compressive pressure, as shown in Figure 17.

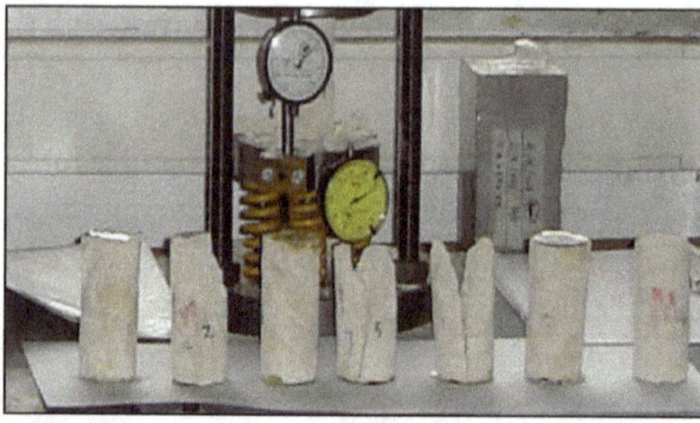

(a)

(b)

(c)

Figure 17. Rock specimens under investigation, after uniaxial creep test. (a) Calcarenitic rock specimens, Necropolis of Mustafa Kamel. (b) Oolitic intraclastic limestone specimens, El-Shatby Necropolis. (c) Sandy oolitic limestone specimens, Catacombs of Kom El-Shoqafa.

7.9 Triaxial creep tests on the calcarenitic and sandy oolitic rock specimens under investigation

The purpose of triaxial creep tests is to determine the viscosity and plastic parameters of the soft rock samples under confined conditions and to investigate the effects of axial stress and fortified pressure. Time-related parameters are monitored, recorded, and analyzed.

7.10 Test methods

Two samples of rock (length = 91–103 mm, diameter = 41–44 mm) were tested from each site under different constant axial pressures and different static pressure pressures for approximately 300 h. The experimental procedure follows the ASTM standard (ASTM D4406-93). The compression machine (fusion machine, 5000 kN) is used to apply the fixed axial load to the samples. Rock samples were placed in a three-axis cell (GDS) to provide constant confining pressure (Figure 18). The collected sample (Test # 1) of Catacomb of Kom El-Shoqafa is immediately loaded to the axial stress required at 1.45 MPa to limit the pressure by 225 kPa, and the applied axial stress was adjusted twice: the initial applied pressure was increased to 1 = 2, 17 MPa (+50%) after 98 h (2) then increased to 1 = 2.53 MPa (+74%) after 125 h (3). The number in brackets refers to Figure 18, which displays the strain versus the time curve. Where the axial stress of up to 1.45 MPa and inventory pressures 510 kPa. Axial stress was not adjusted until the test ends after 200 h with steady-state creep with a small stress rate and without sample failure.

Samples collected (test # 1) from the Shatby Necropolis site were immediately loaded on the required axial stress at 2.63 MPa to limit pressure at 210 kPa, the applied axial stress was adjusted once: the initial applied pressure was increased to = 1 = 3.30 MPa (+25%) after 26 h (2).

In Mustafa Kamel Test No. 2, the sample was loaded on the axial stress required at 2 MPa to limit pressures of 600 kPa without modifying the axial stress until the end of the test at 300 h without high creep.

During testing, axial distortion and time are recorded. The frequency of reading is once every second at the beginning of the test, and gradually decreases to once every half an hour after the first day of the test. This also depends on the deformation rate of each sample. The results are presented by stress time curves in Figures 19–24. Axial stress and axial pressure values are calculated.

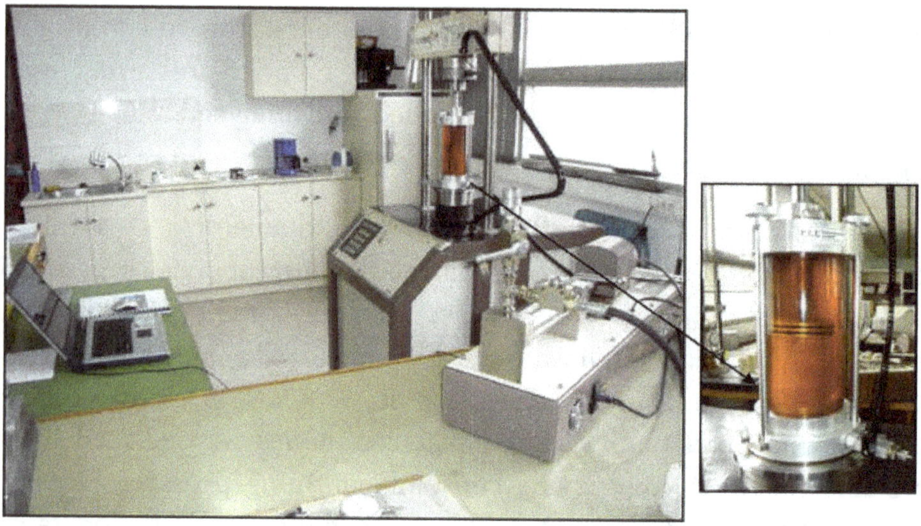

Figure 18. Triaxial creep test device, with constant axial load under confining pressure. Triaxial creep test device. The cylindrical specimen placed inside (GDS) cell is loaded vertically using the compression machine.

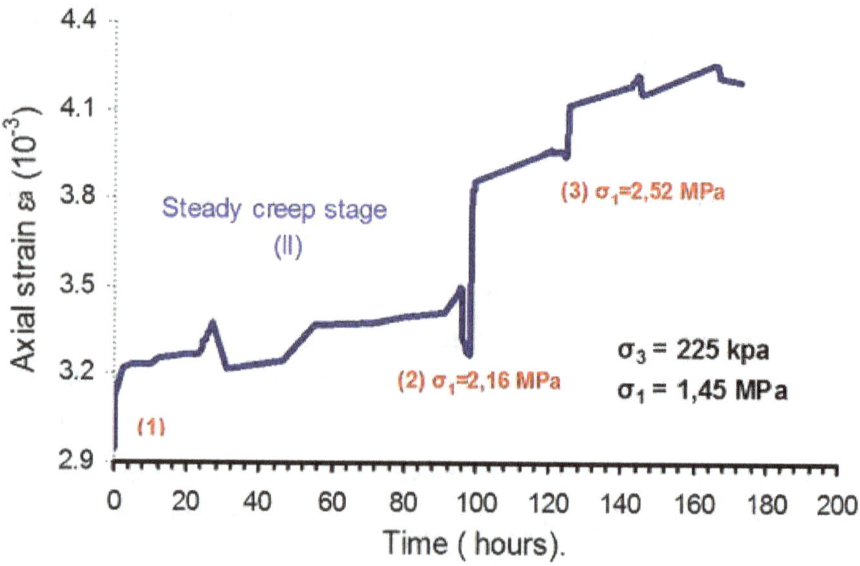

Figure 19. Strain versus time during the catacomb of Kom El-Shoqafa, triaxial creep test no. 1.

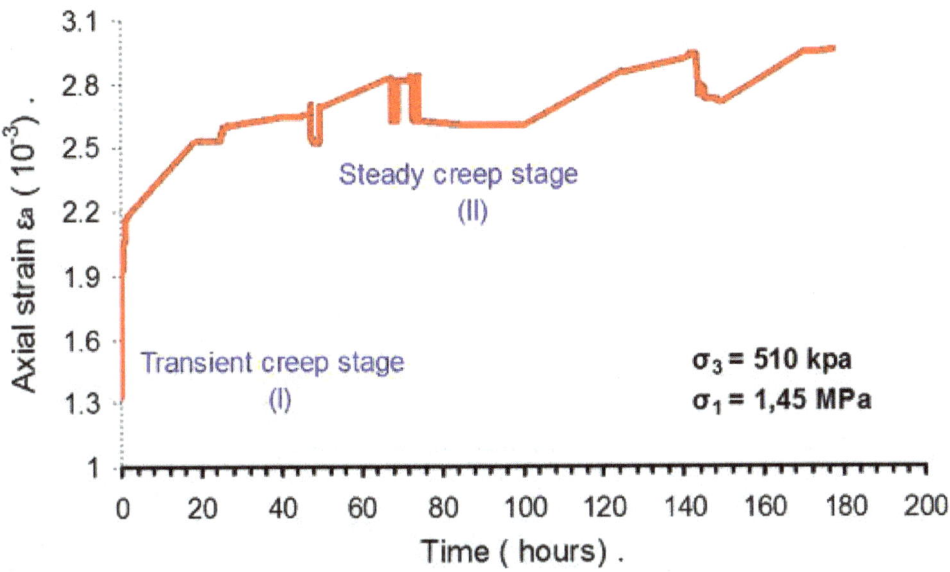

Figure 20. Strain versus time during the catacomb of Kom El-Shoqafa, triaxial creep test no. 2.

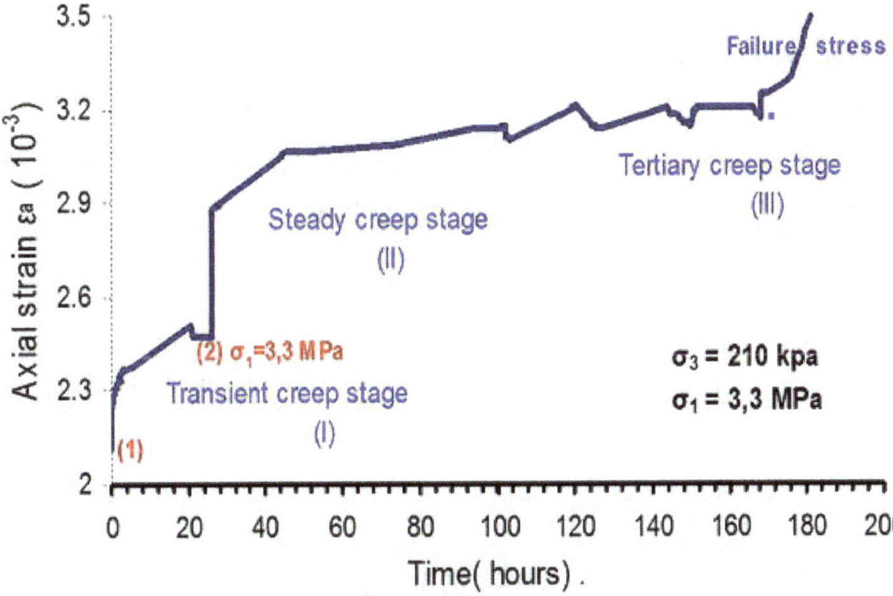

Figure 21. Strain versus time curve during El-Shatby Necropolis, triaxial creep test no. 1.

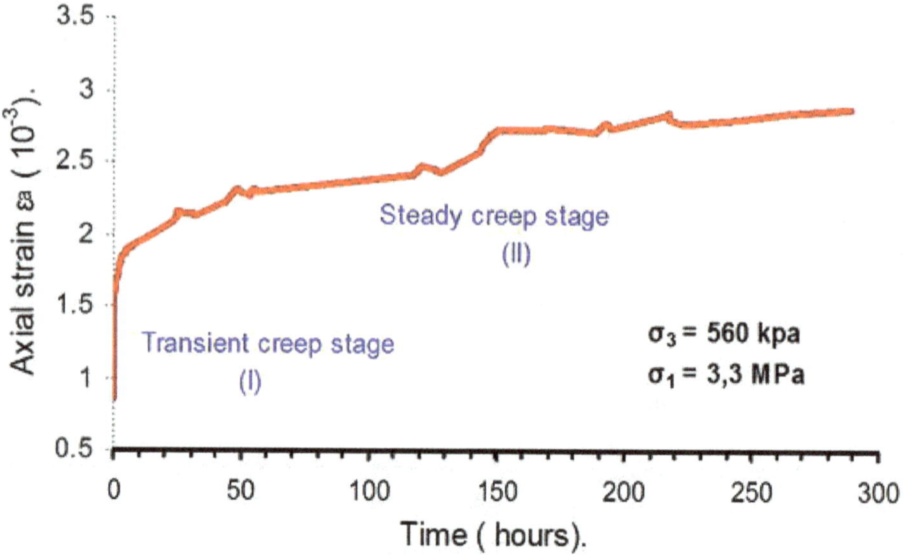

Figure 22. Strain versus time during El-Shatby Necropolis, triaxial creep test no. 2.

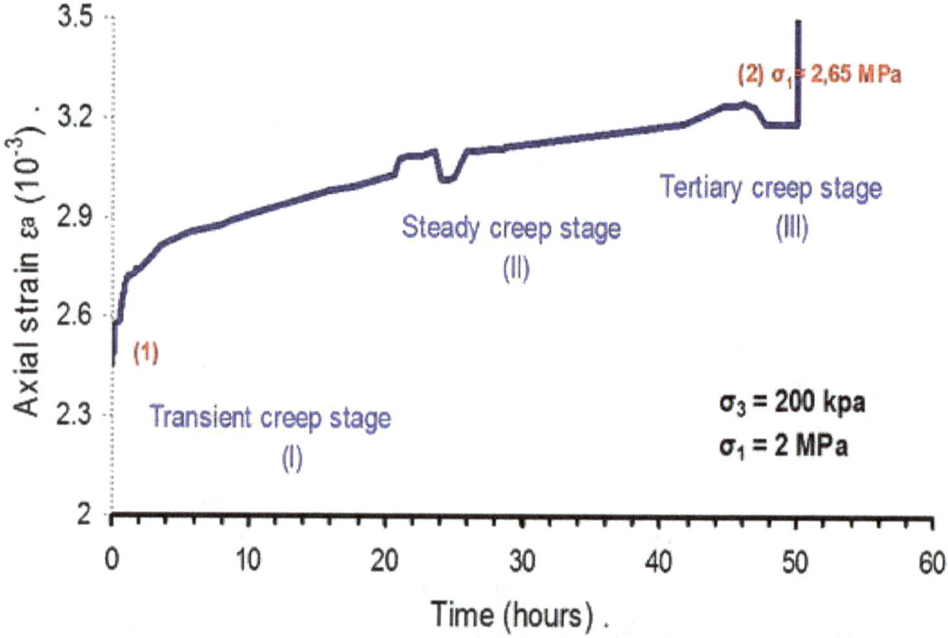

Figure 23. Strain versus time during Mustafa Kamel Necropolis, triaxial creep test no. 1.

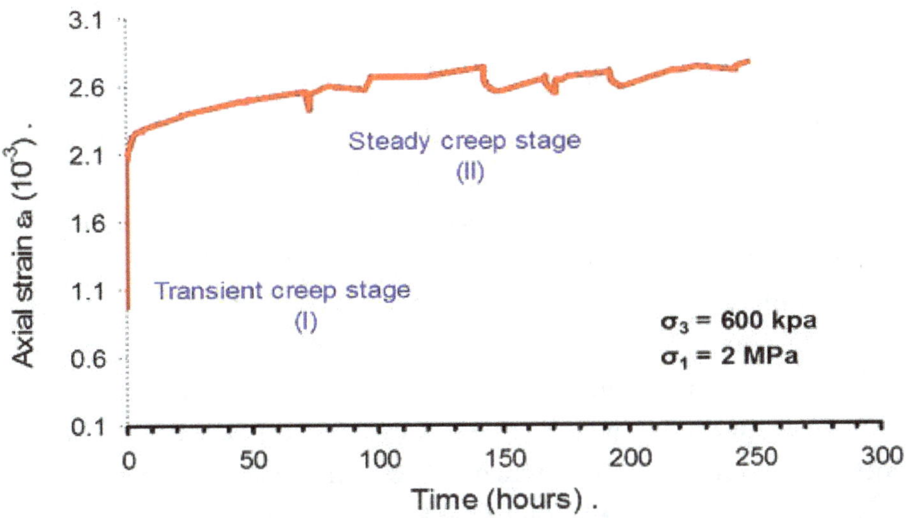

Figure 24. Strain versus time curve during Mustafa Kamel Necropolis, triaxial creep test no. 2.

8. TEST RESULTS

8.1 Catacomb of Kom El-Shoqafa site

Axial strain time curves are shown in shapes (Figures 19 and 20). The curves represent transient and transient creep conditions of rock samples under constant axial load and compression pressure. Instantaneous strains were observed immediately after loading the range from 3 × 10–3 to 3.2 × 10–3 for the test number_1, and 1.3 × 10–3 to 2.2 × 10–3 for the test number_2. All samples show a long "slow low" primary transient creep and steady-state creep stages until the end of the test without acceleration or triple creep resulting in sudden failure. Observations on subsequent tests show that deformation increases rapidly at first to the first few hours of testing and tends to remain constant after that. Stress rates in a steady state are 0.01 to 0.02 × 10–3 h–1.

In this test, it was observed that crawling during the catacomb of Kom El-Shoqafa site Test no_1 (where = 3 = 225 kPa) was faster than crawling during the catacomb of Kom El-Shoqafa site test no_2 (where = 3 = 510 kPa) with the same axial pressure 1 = 1.45 MPa: From 1 to 96 h after the start of each test, the axial strain accumulated (10–3) 3.5 for the catacomb of Kom El-Shoqafa site Test No. 1 and 2.7 for the catacomb of Kom Shoqafa website Test No. 2.

8.2 El-Shatby Necropolis

Axial strain time curves are shown in shapes (Figures 21 and 22). The curves represent temporary and transient creeps of rock samples under constant axial load and confined pressure. Instantaneous strains were observed immediately after the loading range from 2.5 × 10−3 to 2.9 × 10−3 for number_1 test, and from 0.91 × 10−3 to 1.8 × 10−3 for number_2 test. All samples show a long "slow low" primary transient creep and steady creep stages (constant slope) up to the end of the test at 300 h except test number_1, which showed acceleration or triple creep stage leading to a sudden sample failure at 180 h where the confined pressure 3 was small, that is, 210 KPa, while at test number_2, it was 560 kPa and the axial pressure was the same for the eyes 1 = 3.3 Mpa. Observations on subsequent tests showed deformation increases rapidly at first to the first few hours of testing and tends to remain constant after that. The first sample failed after the end of the test. Pressure rates in the steady state are 0.01 to 0.015 × 10−3 h−1.

It was observed that creep through Shatby site Necropolis, test number 1 (where k3 = 210 kPa) was faster than creep through Shatby cemetery site, test no_2 (where = 3 = 560 kPa) under the same axial pressure 1 = 3.3 MPa: From 1 to 150 h after the start of each test, the accumulated axial strain (10−3) was 3.2 for Shatby Necropolis site Test site 1 and 2.4 for Shatby site Necropolis Test No. 2.

8.3 Mustafa Kamel Necropolis

Table 3 summarizes the results of the triple axial crawl test. Axial strain time curves are shown in shapes (Figures 23 and 24). The curves represent transient and transient creep conditions of rock samples under constant axial load and compression pressure. Strains observed immediately after the loading range from 2.5 × 10−3 to 2.7 × 10−3 for test number_1, and from 0.98 × 10−3 to 2.3 × 10−3 for test number 2. All samples show a long initial transient creep "characterized by slow rate of decline" and steady creep phases (constant slope) up to the end of the test at 300 h except the first sample, which shows a transient, steady, and triple-accelerated creep phase leading to sudden sample failure at 49 h immediately after adjusting the axial pressure from 1 = 2 MPa to 1 = 2.65 MPa. Observations on subsequent tests showed deformation increases rapidly at first to the first few hours of testing and tends to remain constant after that. Pressure rates in the steady state are 0.01 to 0.015 × 10−3 h−1.

Table 3. Triaxial creep test, testing program.

Specimen No.	Testing period	Confining pressure ($_3$)	Time (h) 1 3 26 51 52 98 125 200 300		
Catacomb of Kom El-Shoqafa test _1 (Sandy oolitic limestone)	From **19/10/2016** to **27/10/2016**	**225 KPa**	**1= 1.45 MPa**	**1= 2.17 MPa**	**1= 2.53 MPa**
Catacomb of Kom El-Shoqafa test _2	From **21/11/2016** to **28/11/2016**	**510 KPa**	**1= 1.4 MPa**		
El-Shatby Necropolis test _1 (oolitic intraclastic limestone)	From **9/11/2016** to **16/11/2016**	**210 KPa**	**1= 2.63 MPa**	**1= 3.30 MPa**	
El-Shatby Necropolis test _2 (oolitic intraclastic limestone)	From **29/11/2016** to **11/12/2016**	**560 KPa**	**1= 3.31 MPa**		
Mustafa Kamel Necropolis test _1 (Calcarenite rock)	From **16/10/2016** to **18/10/2016**	**200 KPa**	**1= 1.32 MPa**	**1= 1.98 MPa**	**1= 2.6 MPa**
Mustafa Kamel Necropolis test _2 (Calcarenite rock)	From **12/12/2016** to **22/12/2016**	**600 KPa**	**1= 2 MPa**		

In this test, it was observed that crawling through the site of Mustafa Kamel's cemetery in test 1 (where = 3 = 200 kPa) was faster than crawling through the site of Mustafa Kamel's cemetery. No_2 test (where = 3 = 600 kPa) under the same axial pressure = 1 = 2 MPa: from 1 to 45 h after the start of each test, the accumulated axial strain (10−3) was 3.3 for Necropolis of Mustafa Kamel site Test no. 1 and 2.5 for the site of Mustafa Kamel cemetery test site 2.

Thus, the prevalence of cracking (in the fragile field) and pore breakdown (under high pressure conditions) are the prevailing deformation mechanisms of the selected rocks.

The cumulative results of various three-axis crawl tests, conducted at tight pressures ranging from 200 to 600 kPa, showed that crawling reduces the level of brittle stress on failure by 15–20% in relation to standard tests, and similarly, the resulting stress threshold (e.g., Pore breakdown (reduction)) is reduced by the same amount, while the volumetric component of the strain is diluted only in the absence of confined pressure, and shrinks completely even when 3 decreases.

The instantaneous creep strain depends on axial pressure and confining pressure. In general, increased continuous axial pressure leads to greater axial stress. The pressure rate under high axial pressure is greater than the pressure under the lower axial pressure for the same fixed pressure. The higher the confined pressure, the smaller the resulting pressure. Comparison of results obtained from other soft rocks/ salts indicates that the stress rate depends on the stress and previous strain. This is also consistent with the conclusion of Courthouse and Ong et al. who describe soft rocks as close.

The time-based foundational model of soft rocks developed by Zhang et al. can reproduce the general crawl characteristics of soft rocks with high precision. The crawl failure time to load the strain of the aircraft is longer than that of the three-axis load because the strain load frame controls the sample to expand.

9. MICROMECHANICS OF CREEP IN THE CALCARENITIC ROCKS

There is now a large body of evidence that rock deformation at low temperatures and pressures occurs through two mechanisms widely referred to as faulty flow and ductile flow. The term ductile is often used in three different contexts, including (1) plastic deformation of single crystals, (2) homogeneous deformation or uniform flow, and (3) deformation over a certain amount of stress. Here we will use the term ductile in the macroscopic sense of homogeneous deformation where inferior microscopic processes include improved shear pressure, granulation, and granular flow.

Macroscopically, these microscopic processes form the flow of the calcite. Experimental evidence of Caracola flow includes (1) a broad shear area indicating distributed damage and intact granules in an extraction; (2) a large pore breakdown, often accompanied by small intracranial cracks caused by "fragmentation"; and (3) fractures. Unlike this distributed pervasive flow, the standard fragile deformation at low-effective pressures is characterized by an expansive fine fracture, leading to shear localization along narrower fracture zones, which often consist of sections linked to a zigzag pattern (e.g., [44, 45, 46, 47, 48]). In thin sections, the fragile fracture is evidenced by the presence of almost abundant small cracks, away from the shear fracture. Many of these miniature cracks are parallel to the main baseline pressure and may arise from axial splitting of healthy grains [42, 43] or cracking of grain boundaries.

The researchers concurred that the dominant failure mechanisms in highly porous rocks, particularly at high effective medium pressures, are distributed sedimentary rocks, particularly calcarenitic sedimentary rocks, as evidenced by previous experimental investigations [49, 50, 51, 52]. Conversely, the fragile local fracture is the dominant feature in rocks with low porosity and high-porosity rocks with low effective pressure.

10. CONCISE SUMMARY

The catacombs of the Kom El-Shoqafa and Amod El-Sawari (Pompeii's pillar) site, which are situated in the city center and 2.5 km from the sea littoral, are carved into the initial sandy limestone (cement limestone). The lower portions of the cross joints are filled with fragmented sand and saturated with water. Loose sandstone is employed to illustrate this unit. It is medium brown in color and is used to embellish granulated limestone that has been saturated with groundwater. It surpasses the formation of the hayf (Pliocene) or the elder myosin. Actual contact is obscured by surface quadruple deposits. Shatby Cemeteries and Mustafa Kamel Cemeteries, the other two archeological sites situated in close proximity to Alexandria's waterfront, were excavated in internal limestone or calcite (coastalhills). Yellowish white upward transitions to yellow brown at the base. The UCS values obtained from air-dried samples prepared in the vertical direction suggest that the classification adopted by the London Geological Society, which is based on the unrestricted compressive strength and the classification proposed by [33, 50], is accurate. The calcarenitic rocks that are excavated underground are classified as soft to very feeble. In addition, it is in good accordance with the Rock Quality Assignment System (RQD) for these soft rocks, which have a very low quality range from 0 to 25 and an RQD of 15–20% and RR of 18. Furthermore, the static deformation tests results suggest that the rock types in issue exhibit significant deformation. It is important to mention that the silica content at the Catacomb site in Kom El-Shoqafa is higher than in any other area of Alexandria. This may be attributed to sedimentation processes, such as the high silica content that is present in the form of sand grains and does not contain cement. The strength of rock against salt crystallization and moisture pressures within rock cavities is diminished in low rock durability and stiffness due to the presence of a high sand-like grain content. This is due to the fact that it is a sparse rock, in addition to the high concentration of silica granules. It is widely recognized that this particular variety of limestone is distinguished by its inadequate durability. The uniaxial and triple-axis crawling tests can be used to identify three stages of crawling behavior. In certain instances, the primary creep trajectory approaches a constant stress rate known as secondary creep. In high-stress specimens, secondary crawl may manifest as higher creep, which is defined by an elevated stress rate until crawl failure occurs abruptly. In the final two stages, thin vertical cracking commences, followed by hardening. Large fractures spread rapidly and result in a sudden collapse only when the structure is near failure. Long-term experiments were conducted on a secondary creep sample, which demonstrated even strength at 40% of the estimated strength.

The weathering process is linked to structural properties, including the presence of soluble ions in the porous system, the presence of carbon chemical composition, the marine climate with characteristic humidity, and marine spray, as well as groundwater.

AT A GLANCE

Mainly excavated in the calcarenitic limestone formations, the Greek-Roman rock-cut tombs in Alexandria, Egypt, exhibit varying degrees of harm to the rock pillars and ceilings. In order to comprehend the long-term rock mass behavior in specific chambers and its influence on past failures and current stability, uniaxial and triaxial Creep tests and rock mass quality assessments were implemented. The creep behavior of rock is particularly significant in subterranean works, particularly for archeological structures that are subjected to significant initial stresses. These conditions result in nonreversible deviatoric creep strains that accumulate over time under constant stress. Several methodologies have been developed to characterize the time-dependent deformation, including analytical, empirical, and numerical methods. Our analyses indicate that the Roman monuments at Alexandria have been divided into rock masses of inferior quality. Tension fractures and joints, which were persistent discontinuities that were unfavorably oriented, frequently facilitated the failure of ceilings and pillars of rock. However, additional failures were associated with the disintegration of calcarenitic and oolitic limestones. Our findings indicate that the design of monumental tombs during the Roman era was altered to mitigate the risk of rock falls and prevent collapses due to the presence of low-strength granite masses.

REFERENCES

1. Berest P, Charpentier JP, Blum PA. Creep of rock under small loading. In: Proceedings of SMRI Fall Meeting; Washington DC; 1999. pp. 11-24
2. Costin LS. Time-dependent deformation and failure. In: Atkinson BK, editor. Fracture Mechanics of Rock. San Diego, CA: Academic; 1987. pp. 167-215

3. Das S, Scholz CH. Theory of time-dependent rupture. Journal of Geophysical Research. 1981;86:6039-6051

4. De Meer A, Spiers CJ. Creep of wet gypsum aggregates under hydrostatic loading conditions. Tectonophysics. 1995;245:171-183

5. ASTM D 4406–93 (Reapproved 1998). Standard test method for creep of cylindrical rock core specimens in triaxial compression; 1998. pp. 1-5

6. Aydan IT, Zbay U, Kwasniewski M, Shariar K, Okuno T, Zgenoglu A, et al. ISRM suggested methods for determining the creep characteristics of rock. Rock Mechanics and Rock Engineering. 2014;47:275-290

7. Bieniawski ZT. Mechanism of brittle fracture of rock. International Journal of Rock Mechanics and Mining Sciences. 1967;4:395-430

8. Dusseault MB, Fordham CJ. Time dependent behaviour of rocks. Comprehensive Rock Engineering. 1993;3:119-149

9. Hemeda S. Engineering failure analysis and design of support system for ancient Egyptian monuments in Valley of the Kings, Luxor, Egypt. Geoenvironmental Disasters. 2018;5:12. DOI: 10.1186/s40677-018-0100-x

10. Haupt M, Natau O. Uniaxial relaxation test on rock salt. In: Proceedings of the 2nd Conference on Mechanical Behaviour of Salt; Hanover; 1984. pp. 180-185

11. Ladanyi B. Time dependent response of rock around tunnels. Comprehensive Rock Engineering. 1993;2:77-112

12. Lama RD, Vutukuri VS. Handbook on Mechanical Properties of Rocks-Testing Techniques and Results. Vol. 21978. pp. 209-323

13. Maranini E, Brignoil M. Technical note: Creep behaviour of a weak rock: Experimental characterization. International Journal of Rock Mechanics and Mining Sciences. 1999;36:127-138

14. Nomikos. Supported axisymmetric tunnel within linear viscoelastic burger rocks. Rock Mechanics and Rock Engineering. 2011;44:553-564

15. Pan YW, Dong JJ. Time-dependent tunnel convergence I. Formulation of the model. International Journal of Rock Mechanics and Mining Science and Geomechanics Abstracts. 1991;28:477-488

16. Robertson EC. Creep of Solenhofen limestone under moderate hydrostatic pressure. Geological Society of America Memoirs. 1963;79:227-244

17. Wang Y, Qi J, Yang C, Wei J. A study of nonlinear creep law in deep rocks. Rock and Soil Mechanics. 2005;26(1):117-121

18. Zhu H, Ye B. Experimental study on properties of rock creep in saturation. Chinese Journal of Rock Mechanics and Engineering. 2002;21(12):1791-1796

19. Zhang Z, Luo J. Study on creep properties of rock under step load. Chinese Journal of Rock Mechanics and Engineering. 2004;23(2):218-222

20. Liu J, Yang C, Li X, Jiang D. Testing study on creep of coal rocks in the tunnel of Wankai speedway. Chinese Journal of Rock Mechanics and Engineering. 2004;23(22):3794-3798

21. Parkin AK. Creep of rockfill (Part A). In: Maranhadas Neves E, editor. Advances in Rockfill Structure. London: Kluwer Academic Publishers; 1992. pp. 221-239

22. Shen Z, Zhao K. Back analysis of creep deformation of rockfill dams. Journal of Hydraulic Engineering. 1998;6:1-6

23. Guo X, Wang D, Cai X, Dong L. Rheological analysis of concrete faced rock-fill dam. Journal of Hydraulic Engineering. 1999;11:42-46

24. Wang M, He X, Cheng Z. Current situation and prospect of studies on rheology property of course stuff. Rock and Soil Mechanics. 2003;Suppl:451-454

25. Liu Y, Li Y, Sun M. A new method of permeability test for loose rock. Ground Pressure and Strata Control. 2002;19(4):108-110

26. Hemeda S. An integrated approach for the pathology assessment and protection of underground monuments in seismic regions.

Application on some Greek-Roman Monuments in Alexandria, Egypt [PhD thesis]. Greece: Aristotle University of Thessaloniki; 2008

27. Hemeda S, Pitlakis K. Serapeum temple and the ancient annex daughter library in Alexandria, Egypt: Geotechnical-geophysical investigations and stability analysis under static and seismic conditions. Engineering Geology. 2010;113:33-43

28. Hemeda S, Pitilakis K, Bandis S, Papayianni I, Gamal M. The Underground monuments (Catacombs) in Alexandria, Egypt. In: Proceedings of the 4th International Conference on Earthquake Geotechnical Engineering; Thessaloniki, Greece; 25–27 June, 2007. pp. 715-738

29. Bieniawski ZT. The geomechanics classification in rock engineering applications, In: Proceedings of the Fourth International Conference on Rock Mechanics, Vol. 2. Montreox: ISRM; 1979. pp. 41-48

30. Aydan O, Akagi T, Ito T, Kawamoto T. Prediction of behaviour of tunnels in squeezing ground. Journal of JSCE. 1992;440(3–9):73-82

31. Aydan O, Akagi T, Kawamoto T. The squeezing potential of rocks around tunnels: Theory and prediction. Rock Mechanics and Rock Engineering. 1993;26(2):137-163

32. Aydan O, Kawamoto T. The stability assessment of a large underground opening at great depth. In: 17th International Mining Congress and Exhibition of Turkey (IMCET 2001); Ankara, Vol. 1; 2001. pp. 277-288

33. Aydan O, Ulusay R. Geotechnical and geo-environmental characteristics of man-made underground structures in Cappadocia, Turkey. Engineering Geology. 2003;69:245-272

34. Aydan O, Ulusay R, Kawamoto T. Assessment of rock mass strength for underground excavations. In: The 36th US Rock Mechanics Symposium; 1997. pp. 777-786

35. Deere DU, Varde OA. General report, engineering geological problems related to foundations and excavations in weak rocks.

In: Proceedings of the 5th International Association of Engineering Geology Congress, 1986. Vol. 4. Balkema; 1990. pp. 2503-2518

36. Hemeda S. 3D finite element coupled analysis model for geotechnical and complex structural problems of historic masonry structures: conservation of Abu Serga church, Cairo, Egypt. Heritage Science. 2019;7:6. DOI: 10.1186/s40494-019-0248-z

37. Hemeda S. Geotechnical and geophysical investigation techniques in Ben Ezra Synagogue in Old Cairo area, Egypt. Heritage Science. 2019;7:23. DOI: 10.1186/s40494-019-0265-y

38. Hemeda S, Pitilakis K. Geophysical Investigations at Cairo's Oldest, the Church of Abu Serga (St. Sergius), Cairo, Egypt. Research in Nondestructive Evaluation. 2017;28(3):123-149. DOI: 10.1080/09349.847.2016.11439.91

39. Bukovansky M, Richard DP, Week KR. Influence of slope deformations on the tombs in the valley of the kings, Egypt. In: Engineering Geology and the Environment-Proceedings-Symposium; Athens, Vol. 3; 1997. pp. 3077-3080

40. Berest P, Charpentier JP, Blum PA, Gharbi H. Very slow creep tests on rock samples. International Journal of Rock Mechanics and Mining Sciences. 2005;42:569-576

41. Brace WF, Kohlstedt DL. Limits of lithospheric stress imposed by laboratory experiments. Journal of Geophysical Research. 1980;85:6348-6352

42. Brace WF, Paulding BW, Scholz CH. Dilatancy in the fracture of crystalline rocks. Journal of Geophysical Research. 1958;71:3939-3953

43. Charles RJ. Fatigue of glass, I. Journal of Applied Physics. 1966;29:1549-1560

44. Adachi T, Oka F. An elasto-plastic constitutive model for soft rock with strain softening. International Journal for Numerical and Analytical Methods in Geomechanics. 1995;19:233-247

45. Arces M, Nocilla, Aversa S, Lo Cicero G. Geological and geotechnical features of the "Calcarenite di Marsala". In: Evangelista, Picarelli, editors. Proceedings of the Geotechnics of Hard Soils-Soft Rocks. Rotterdam: Balkema; 1998

46. Atkinson BK. Introduction to fracture mechanics and its geophysical applications. In: Atkinson BK, editor. Fracture Mechanics of Rocks. San Diego, CA: Academic; 1987. pp. 1-26

47. Atkinson BK, Meredith PG. The theory of sub-critical crack growth with applications to minerals and rocks. In: Atkinson BK, editor. Fracture Mechanics of Rocks. San Diego, CA: Academic; 1987. pp. 111-166

48. Aversa S. Mechanical behaviour of soft rock: Some remarks. In: Proceedings of the Workshop on "Experimental Characterization and Modelling of Soil and Soft Rocks. Napoli; 1991. pp. 191-223

49. Aydan O, Tokashiki N, Kawamoto T. Microstructure models for porous rocks to jointed rock masses. In: Third Asia-Pacific Conference on Computational Mechanics; Seoul, Vol. 3; 1996. pp. 2235-2242

50. Cecconi M, Viggiani G. Physical and structural properties of a pyroclastic soft rock. In: Evangelista, Picarelli, editors. Proceedings of the Geotechnics of Hard Soils-Soft Rocks. Rotterdam: Balkema; 1998

51. Cooper MR. The mechanics of uncemented carbonate sands. Geotechnique. 1990;40(4):607-627

52. Cripps JC, Coulthard JM, Forster A, Hencher SR, Moon CF. The engineering geology of weak rock. In: Proceedings of the 26th Annual Conference of the Engineering Group of the Geological Society, Leeeds (AA Balkema, Engineering Geology Special Publication, 8); 1993. 510pp

03 MODELS OF HOW THE GROUND REACTS TO A STEEL-COMPLIANT RISER IN DEEPWATER CLAYS

1. INTRODUCTION: AN OVERVIEW

Catenary steel compliance risers (SCRs) have been introduced as a new addition to the riser family, using the catenary equation that has been instrumental in constructing bridges worldwide. Silicon-controlled rectifiers (SCRs) are frequently employed in conjunction with tension leg platforms (TLPs), floating production storage and offloading vessels (FPSOs), semisubmersibles, spars, as well as fixed structures, compliant towers, and gravity structures. Since 1994, floating platforms have been used in conjunction with SCR systems. These platforms were initially employed as export risers for the Auger TLP in water depths of 872 meters [1, 2]. Since then, SCRs have been utilized in various applications. The proliferation of SCRs is rapidly growing because to their inherent simplicity, cost-effectiveness, and widely recognized material features. A catenary riser, which is not supported by any structures, is attached to a floating production vessel. The riser is suspended at a specific top angle. The riser is suspended without support and gradually slopes downward to the seabed at the landing point (TDP). The TDP utilizes a trench where the SCR pipe is embedded and then uniformly ascends to the surface, where it remains stationary, functioning as a static pipeline. SCRs can be characterized as comprising three components [3], as depicted in Figure 1:

- Catenary zone, where the riser suspends in a catenary section
- Buried zone, where the riser pipe penetrates into a trench
- Surface zone, where the riser pipeline rests on the seabed.

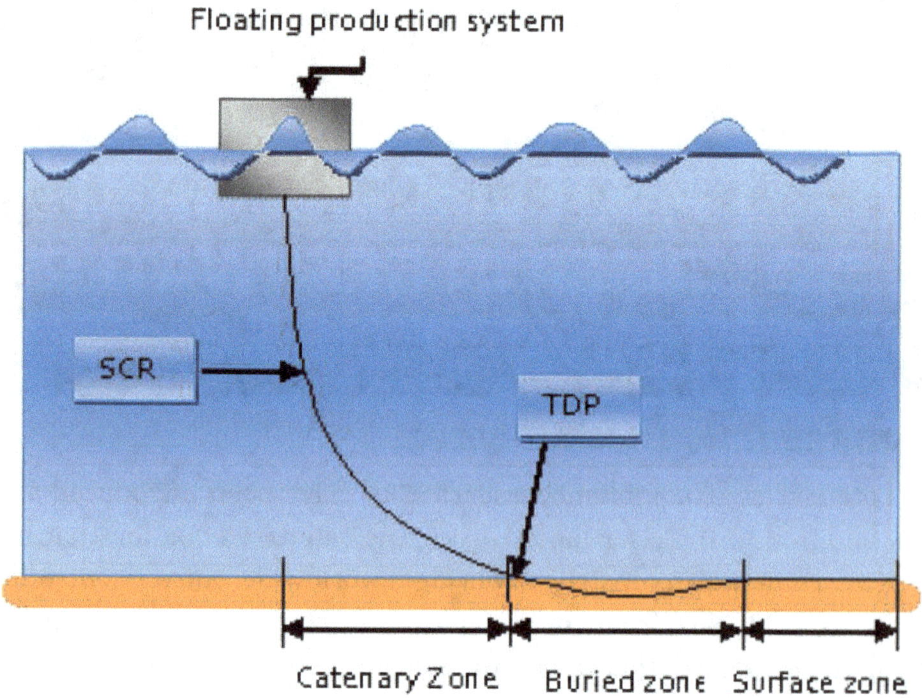

Figure 1. General SCR arrangement.

An intricate interplay occurs between the SCR and seabed when the SCR is subjected to oscillatory movements. The most crucial fatigue hotspot for SCRs is located in the TDZ. The interaction between the SCR and the seafloor is a crucial component that must be taken into account when assessing strength and fatigue. Accurately representing this interaction response remains a challenge and has been a prominent area of focus in academic study. Several studies have been dedicated to comprehending the interaction between soil and risers. To improve the accuracy of predicting the fatigue life of the SCR, it is necessary to have a precise model that characterizes the stiffness of the seabed and a realistic representation of the load/deflection curve. This chapter provides a comprehensive and up-to-date overview of the latest research on models that study the interaction between soil and risers. In this paper, we will provide a concise overview of previous research on the interaction between the seabed and risers, including simulation models and load/deflection models. We will describe and analyze these works in detail.

2. SCR CONFIGURATION DESIGN

The catenary riser length is estimated using simple geometric considerations, as following [4]:

$$L = \left(\frac{D - (MBR)A}{\cos\theta}\right) + (0.5\pi(MBR)A) \quad (1)$$

where L is the total length of the riser, D is the water depth, A is a factor depending on severity of environment (1.0 for mild environments and 1.2 for severe environments), θ is the riser top angle to vertical, typically between 10 and 25 degrees depending on severity of environment and water depth, and MBR is minimum bend radius based on 80% material yield strength. An additional riser pipeline length of approximately 750 m should be included to allow for TDP movement between near and far offset conditions as shown in Figure 2.

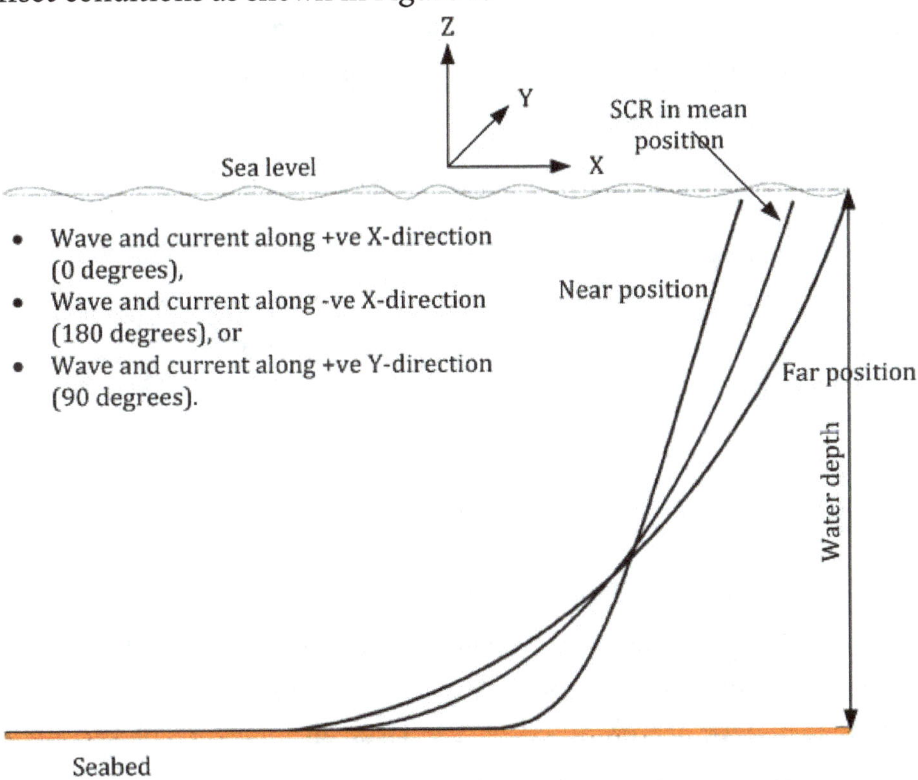

Figure 2. Schematic of SCR configuration and vessel offsets.

The SCR's static configuration must be determined before carrying out the dynamic analysis. The initial stage of any analysis of an SCR is the computation of its configuration under a set of static forces. The catenary equation gives a good first approximation for this, but in their basic formulation it only involves loads due to riser weight and assumes a riser pipe of zero bending stiffness, as presented before. However, the SCR static analysis is a large deflection non-linear behaviour problem with the influences of bending and tensional stiffness included. Therefore, many approaches have been developed to handle this problem using a combination of catenary equations and numerical techniques through iterative analysis. A review of existing approaches can be found in [5]. Figure 3 shows an example of static configuration of an SCR in a 910 m water depth with a hang-off angle of 20° and a 273 mm outer diameter connected to the floating platform (FP) in the zero mean offset position (i.e., the FP is in its initial position without drifting in any direction), which is calculated using OrcaFlex/finite element analysis (FEA) software.

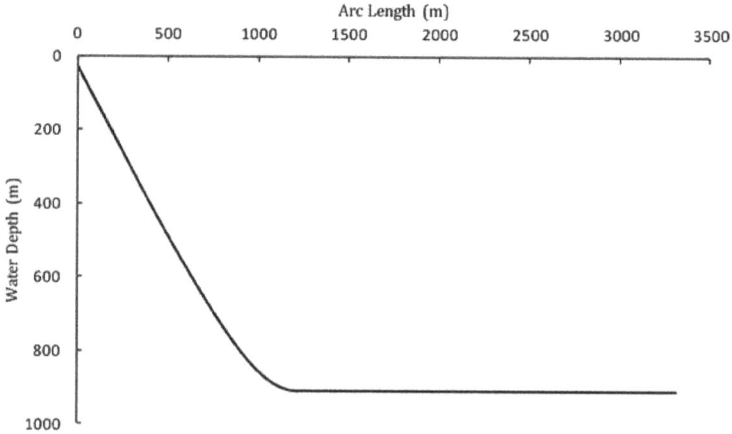

Figure 3. Static configuration of SCR model.

Generally, SCRs have a limited amount of additional pipeline length available to accommodate the FP motions. Alterations in the catenary suspended length are obtained by the riser either being picked up or laid down on the seabed. Limitations are approached when either the SCR tension at the FP becomes too great as the FP drifts away from the TDP (far load case, as shown before in Figure 2) or when the bending stresses near the seabed become too great as the FP drifts towards the touchdown point (near load case). SCRs are less appropriate for FPSO applications where vessel offsets are considerably higher. Figure 4 shows the effect of the horizontal vessel offset on the horizontal projection of the TDP. While the top of the SCR has the highest tension and lowest bending moment, the TDP has the lowest tension and the highest bending moment. The maximum bending stress and effective tension along the SCRs' arc length and the horizontal projection of the TDP due to the vessel offsets are presented in Figures 5 and 6, respectively.

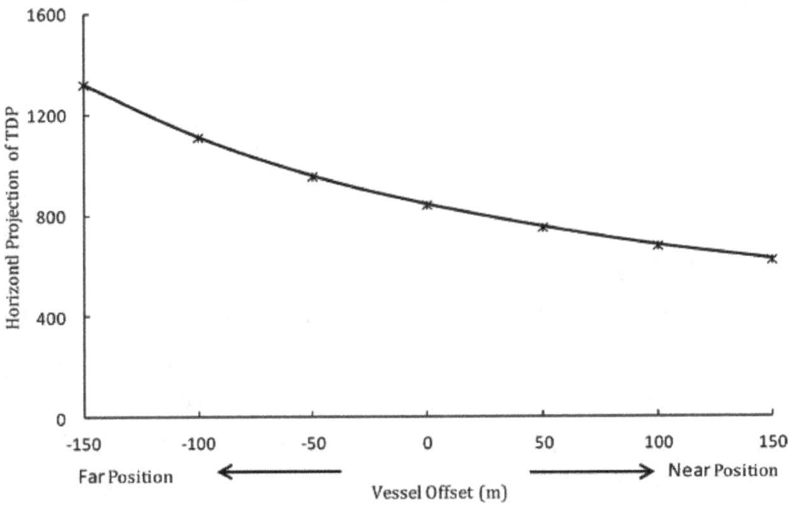

Figure 4. Effect of the vessel offsets on the horizontal projection of the TDP.

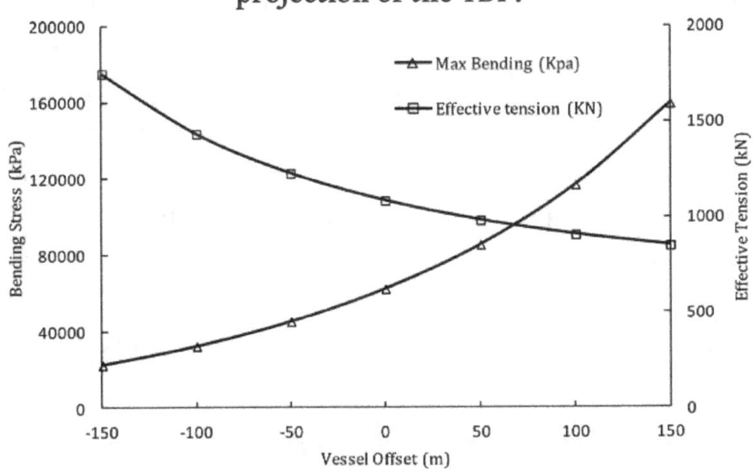

Figure 5. Alterations of maximum bending stress and maximum effective tension with the horizontal vessel offset.

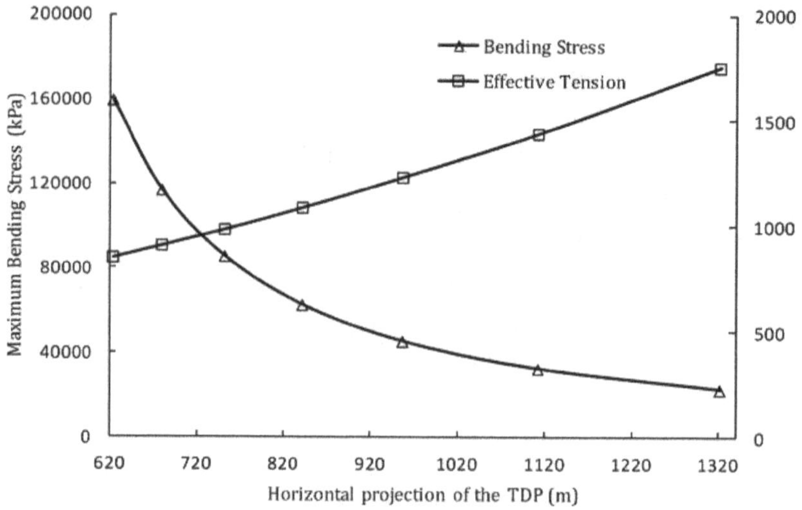

Figure 6. Alterations of maximum bending stress and effective tension with the horizontal projection of the TDP.

The vessel offset governs the maximum bending stress at the TDP and also the maximum tension at the riser's top end. In the left region of Figure 6, where the vessel drifts towards the TDP (near load case), the bending stress at the TDP is increased rapidly within small change in the vessel offset. In the intermediate region, the bending stress and tension are slightly increased with the vessel offset. In the right region, where the vessel drifts away from the TDP (far load case), the bending stress slightly decreased, while the tension increased. Therefore, the conclusion from these results is that the vessel mean position should be offsetting the TDP with a roughly distance of 0.75–1.5 of the water depth. Furthermore, the catenary equation is a simple implementation tool to figure out the load distribution, geometric properties and static loads on an SCR. The specialist non-linear/FEA is implemented for the SCR design to tackle the complex nature of non-linear, large deflection behaviour of SCRs and to be post-processed quickly. The evaluation of the forces and behaviour of SCRs in the TDZ need more sophisticated methods.

3. MODELS FOR SEABED RESPONSE

3.1 Problem description of an SCR pipe embedment

SCR pipe penetration is defined as the depth of penetration of the pipe invert (bottom of pipe), relative to the undisturbed seabed as shown in Figure 7. Pipe penetration affects the riser pipe-seabed contact area, which subsequently affects the axial and passive soil resistance against the riser. Consequently, the passive soil resistance influences the lateral breakout force. Heave of seabed soil during embedment increases the local penetration of the SCR pipe by raising the soil surface level against the shoulders of the pipe. The typical geometry of heave produced during vertical embedment of an SCR pipe is such that the nominal penetration is approximately 50% less than the local embedment relative to undisturbed seabed surface [6].

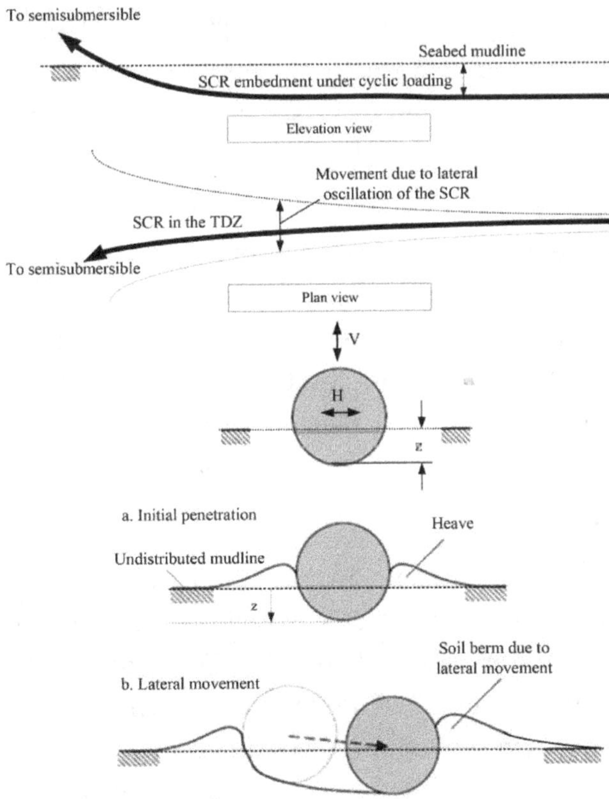

Figure 7. Schematic illustration of the SCR-seabed interaction in the TDZ.

The SCR-seabed interaction response characteristic is a highly non-linear phenomenon. It is important not to restrict the modelling of this interaction to a linear seabed model approximation and the riser analysis techniques must be improved by refining the riser-seabed interaction [7]. SCR-seabed interaction modelling should involve vertical and lateral soil responses to the cyclic loading oscillations of the SCR in the TDZ, which can cause trenching and dynamic embedment of the SCR into the seabed. A typical schematic illustration of the SCR-seabed interaction and trench formation in the TDZ are shown in Figure 7.

3.2 SCR/seabed vertical interaction

The application of SCR systems has increased with the progressive development of hydrocarbon production further offshore and into deeper waters. The SCR-soil interaction at touchdown with the seabed is a major key factor for SCRs. An SCR is subjected to oscillatory motions from the host vessel and wave action. Therefore, the SCR experiences a complex interaction between the riser and seabed in the touchdown area, and deep trenches thus cut into the seabed in the buried zone beyond the TDP [8, 9, 10].

An appropriate SCR-seabed interaction model must be used. The TDZ is one of the key locations where the fatigue damage happens. The sophistication of the interaction model depends on the type of analysis and accuracy required. These interaction models vary from a simple rigid seabed with soil friction coefficients to more sophisticated ones, including vertical and lateral stiffness, friction and suction.

3.2.1 Rigid and elastic seabed

A potential fatigue damage of the SCR in the TDZ is related to maximum bending stress in the SCR, which relies on the soil stiffness of the seabed and the motions of the SCR. For example, the SCR on a soft seabed will have reduced bending stresses when a load is applied, while the one on a rigid seabed will have more critical bending stresses. A rigid surface generally contributes a conservative result, since it is unyielding, while the non-linear soil model is a better approximation of a seabed. Extreme offsets of the floating production unit with soft seabed model may then give higher stresses than those calculated on rigid seabed stiffness, since the catenary pipeline must be broken out of the seabed soil and high suction forces must be overcome. Figure 8 shows a schematic of an SCR close to the TDP with the forces acting on a rigid seabed. The shear force F in the near horizontal segment close to the TDP is given by:

$$F = \frac{dM}{dx} = \frac{d}{dx}\left[EI\left(\frac{d^2y}{dx^2}\right)\right] = EI\left(\frac{d^3y}{dx^3}\right) \quad (2)$$

then d3 y/dx 3 = (w/H)ke-kx and the shear force close to the TDP is thus given by:

$$F = EI\left(\frac{w}{H}\right)ke^{-kx} \quad (3)$$

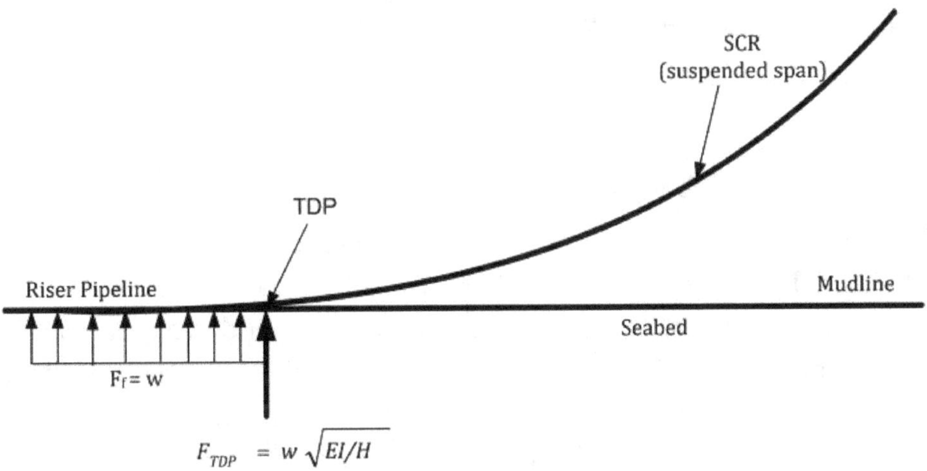

Figure 8. Configuration of SCR close to TDP with a rigid seabed.

The bending moment at the TDP diminishes from the catenary bending moment to zero, and there is a concentrated reaction force. Since $k = \sqrt{H/EI}$, the TDP shear force that is transmitted to the soil [2, 11] is given by:

$$F_{TDP} = w\sqrt{EI/H} \quad (4)$$

where F TDP is the concentrated reaction at the TDP, assuming a rigid surface seabed, F f is the reaction to the pipe resting on the seabed, as shown in Figure 8.

For the elastic soil response, configuration of an SCR close to the TDP is shown in Figure 9 by representing a seabed with a linear elastic model. The curvature in the surface zone (i.e., the pipeline is resting on the seabed) is zero. In the TDZ, the riser's pipe is resting on linear elastic foundations. The solution for a beam element resting on an elastic foundation can be found in [12, 13], who introduced solutions that can be implemented for SCR-seabed interaction.

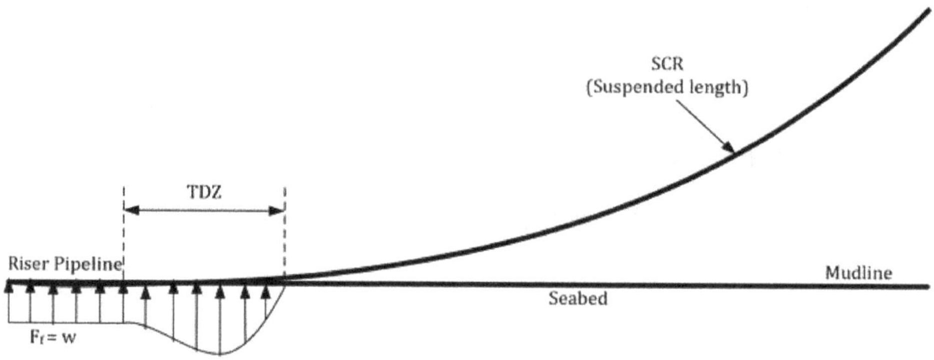

Figure 9. SCR's configuration close to TDZ with linear elastic seabed.

3.2.2 Non-linear load/deflection model

The current practice for the FEA of SCR-seabed interaction response is to model this interaction as structural soil springs [10] by using the developed models for buried pipelines and strip foundation theory. The conventional modelling of riser-seabed interaction use the non-linear elastic load/deflection curves, as described in [14]. Since the resistance force does not exceed the friction resistance limit (V), the soil spring has a constant stiffness coefficient, K. The load/deflection model has zero resistance force at zero displacement, as the pipe displacement is increasing the resistance force also increases linearly until the peak seabed resistance is approached. When the seabed friction exceeds the limit friction force, the resistance force becomes constant with changing pipe displacement (large displacements occur without a further increase in the friction resistance force) and the spring stiffness becomes zero (i.e., slip occurs). The maximum seabed resistance load is given by the backbone curve [15], which corresponds to virgin penetration of the riser pipe into the seabed.

Linear soil stiffness can be used by FEA codes to model the non-linear riser-seabed interaction curves. Linear soil stiffness is defined as the ultimate bearing load divided by the riser pipe displacement, as given below:

$$K = \frac{V}{\Delta} \tag{5}$$

where K is the soil stiffness per unit length; V is the force per unit length; Δ is the riser pipe displacement. Different approaches are used to characterise the linear seabed stiffness, such as secant, tangent and Young's modulus stiffness, for more details see Barltrop et al. [16]. Herein, the secant stiffness type is considered because it is more stable than the tangent stiffness approach and is being used to model the load/deflection curve. The secant stiffness is the average stiffness between two points, typically the origin and the point in question, see Figure 10. Static seabed stiffness is using a linear stiffness to represent the backbone curve in riser-seabed interaction analysis. It is assumed that the riser will penetrate into the seabed until the bearing load equals the submerged riser pipe weight; w = V u where w is the submerged riser pipe weight, and V u is the ultimate bearing load.

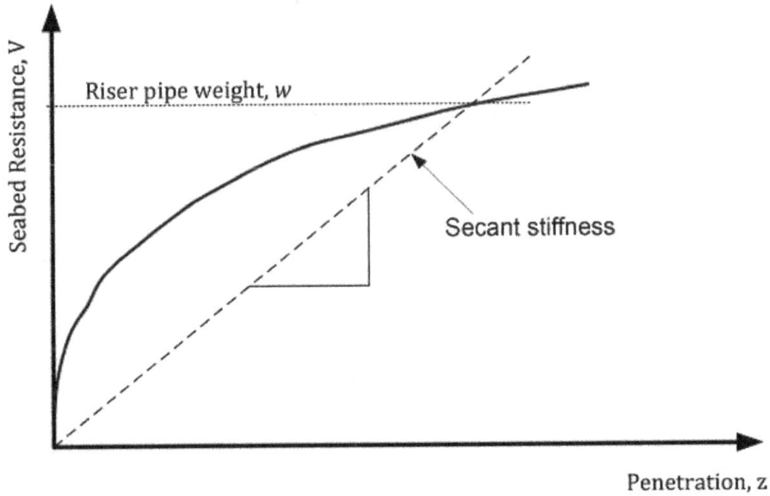

Figure 10. Static and secant stiffness for non-linear seabed V-z model.

The non-linear soil model is recently developed and based on a hyperbolic secant stiffness formulation proposed by Bridge et al. [17],

Aubeny et al. [18], and Randolph and Quiggin [19]. The non-linear seabed model is more sophisticated than the linear model, as it models the non-linear hysteretic behaviour of the seabed in the vertical direction, including modelling of suction effects when the SCR rises up sufficiently. The model uses data such as the pipe diameter, the seabed soil shear strength profile with depth and the soil density as its primary sources. Different functions are used for the initial penetration, for uplift and for re-penetration, whilst the function parameters are updated each time a penetration reversal occurs. This enables the model to capture the hysteretic behaviour of the seabed soil response and the increasing penetration of the pipe under cyclic loading in the vertical plane.

The typical V-z curve patterns [17, 19], as shown in Figure 11, of pipe-soil interaction are produced by laboratory model experiments [15] of vertically loaded horizontal pipes in weak sediment. These curves can be divided into four different paths. The pipe-soil interaction process is described and the depiction of the development of the interaction curve is given in Figure 11, associated with the uplift/re-penetration cycle. If the riser pipe continues to experience oscillatory loading movement, the V-z interaction curve will repeat the loop enclosed by path 1-2-3-1 under the assumption of a non-degradation model.

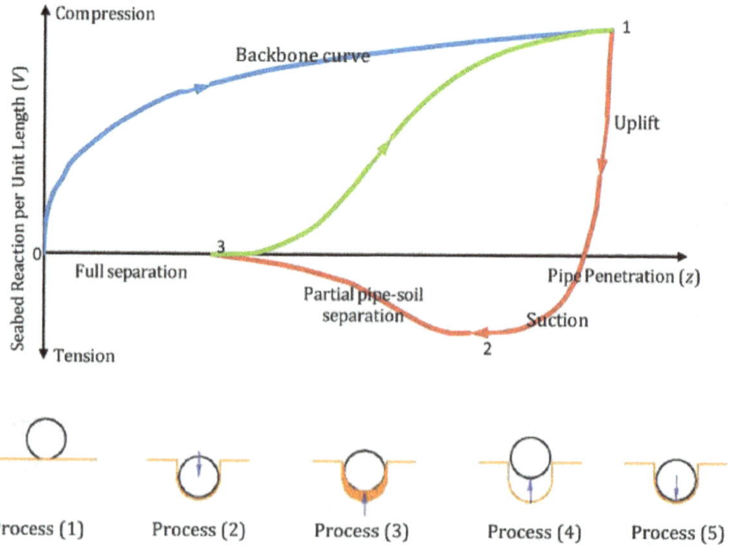

Figure 11. Depiction of typical V-z behavior [7].

3.3 SCR/seabed lateral interaction

One of the main issues encountered with the use of the SCR is the large lateral movements on the seabed due to the FP motions and marine environment. Thus, better understanding of the lateral soil resistance to SCR pipe movements must be considered for SCR design. Many researchers had focused on studying and investigating the pipe-seabed lateral interaction response [6, 20, 21, 22, 23, 24]. Three different approaches [25, 26] can be considered for determining the lateral soil resistance of partially embedded pipelines:

- A single friction factor "Coulomb friction model" approach, where the lateral soil resistance is related to the submerged weight of the pipeline and the soil type. This approach is fairly simplified, as it does not pertain to pipe embedment;
- A two-component model, where the lateral soil resistance consists of a sliding resistance component and a lateral passive pressure component [20, 23, 27]; and
- **A plasticity model approach:** Zhang et al. initially developed the plasticity model for calcareous sand and clays [28]. However, the clay's model is established on the behaviour of shallow flat footings in which a large lateral movement does not occur.

Therefore, the Coulomb friction approach and the two-component soil resistance models for the assessment of SCR global response are presented in this chapter.

3.3.1 Coulomb friction "Bilinear" model

Present industry procedure is to evaluate the soil resistance with a Coulomb friction model, as shown in Figure 12, which expresses the lateral resistance as the product of the effective submerged pipeline vertical force (submerged pipe weight minus hydrodynamic lift force) and a soil friction coefficient that depends solely on soil type. Recommended values of the Coulomb friction coefficient, lie in the range 0.2–0.8, while the displacement to mobilise this resistance is typically 0.1 pipe diameters [20, 23, 24, 29].

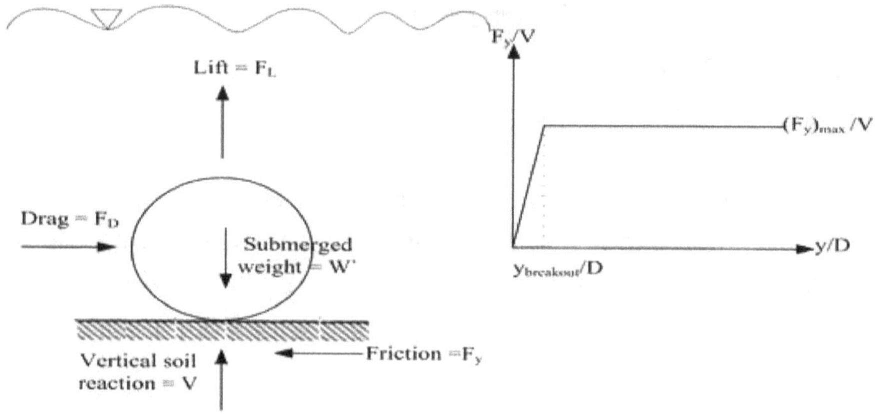

$$(F_y)_{max} = \mu(W' - F_L) \qquad \text{If } (F_y)_{max} \geq F_y = F_D, \text{ sliding will not occur}$$
$$V = W' - F_L \qquad \text{If } (F_y)_{max} < F_D, \text{ sliding will occur}$$

Figure 12. Coulomb friction model analysis.

3.3.2 Tri-linear pipe-seabed interaction model

The experiment results show that the pipe-soil lateral motion is far more complicated than simple coulomb friction. An improved model is essential in order to mimic the effects of soil strength and the load history of the catenary pipeline as well as the associated pipe embedment on the lateral seabed soil resistance. The improved empirical model utilises two components to predict the seabed resistance to lateral pipeline movements, resulting in the improved so-called "two-component model."

The two-component model uses an empirical formula to assess the soil resistance to lateral pipeline motions. The first component depends on the vertical pipe weight (pipe weight minus hydrodynamic lift force) and imitates the sliding resistance of the pipeline along the soil surface, while the second component depends on the pipe penetration and soil strength. Generally, the two-component models are based on empirically fitting laboratory test data. A summary of some of the proposed formulas is given in Table 1. The peak lateral soil resistance is a key parameter for the on-bottom pipeline movement. Several reported methods [20, 23, 27, 30] have been published for the assessment of the lateral soil resistance. These determined resistances were then compared with the results of the available pipe model tests.

Table 1. Lateral resistance models of partially embedded pipelines in soft clay.

Author	Lateral soil resistance formulas	Comments
Wagner et al. [23]	$Fy = \mu(W' - FL) + \beta SuA/D$ Fy = lateral soil resistance = sliding resistance coefficient W' = submerged pipe weight F_L = hydrodynamic lift β = empirical soil passive resistance coefficient Su = undrained shear strength of the clay$A$ = 0.5 × embedded area	Monotonic μ=0.2 β=39.3 Cyclic (oscillations below the monotonic breakout load [<static failure]) Penetration × 2 β=31.4 Cyclic (large displacement oscillations) Penetration × 2.5 β=15.7
Brennodden et al. [20]	$Fy = \mu(W' - FL) + FR$	μ=0.2 FR calculated considering accumulated energy
Verley and Lund [27]	$F_y = F_C + F_R$ $F_y = \mu(W' - F_L) + F_R$ $F_R = 4.13 D S_u \left(\frac{S_u}{\gamma D}\right)^{-0.392} \left(\frac{z}{D}\right)^{1.31}$	Clays (S_u < 70 kPa) μ=0.2
Bruton et al. [30]	$(F_y)_{dimensionless} = \mu v + \frac{3}{\sqrt{\frac{S_u}{\gamma' D}}} \frac{z_{init}}{D}$ $(F_y)_{dimensionless} = \frac{F_y}{S_u D}$ $\frac{z_{init}}{D}$ = normalised initial pipe penetration $v = \frac{V}{DS_u}$ normalised vertical load due to the effective pipe weight $\frac{(f_y)_{res}}{v} = 1 - 0.65\left[1 - \exp\left(-\frac{1}{2}\frac{S_u}{\gamma' D}\right)\right]$	Soft clays (0.15 < S_u < 8 kPa) μ=0.2

Figure 13 shows the lateral load response from step 0 to 3, characterised as follows [31]:

(0–1) First load breakout, with elastic response characterised by the mobilisation displacement and a peak that is dependent on the initial pipe embedment;

(1–2) Suction release phase and elevation correction, depending on initial pipe embedment;

(2–3) Steady state of residual friction.

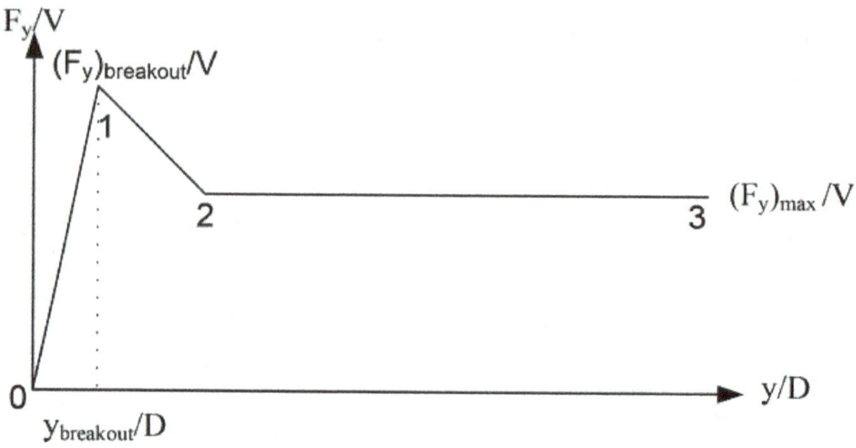

Figure 13. Schematic of the tri-linear soil resistance model.

3.4 SCR/seabed axial interaction

The axial soil resistance for SCR movement is typically modelled using the Coulomb friction model, which is adopted to evaluate the axial resistance of a partially embedded riser pipe, and is expressed as [25, 32]:

$$F_x = \mu_A W_s \tag{6}$$

where F x is the axial soil resistance and AμA is the coefficient of axial coil friction. The typical values for axial friction have been reported to vary between 0.2 and 0.5 for clay soil [33].

4. SUMMARY OF SCR-SEABED INTERACTION MODELS TECHNIQUE

A number of scientific papers have been published on the study of soil-riser interaction. Table 2 details the considerable diversity in the level of sophistication used in the analysis of SCR-seabed interaction response in a representative set of studies published in the last 25 years, with five areas previously highlighted as commonly conservative and broken into components of increasing degree of sophistication and accuracy. Table 2 have been addressed for appropriate modelling of the physical SCR-seabed interaction process and graded with a three levels ranging from α which represents state-of-the-art practice, β a compromise method and γ a conservative method. General and brief discussion on the components presented in Table 2 are as follows:

Table 2. Level of complexity used in SCR-seabed interaction technique.

Author	Year	SCR-seabed interaction models					Pipe-seabed interaction models					Cyclic loading		Wave loading		Analysis	
			Analytical models					Analytical models									
			γ	β	α	α		γ	β	α	α	α	α	β	α	β	α
		Model tests	Rigid seabed	Linear seabed	Non-linear seabed	Trenching effects	Model tests	Rigid seabed	Linear seabed	Non-linear seabed	Trenching effects	Vertical	Lateral	Regular	Irregular	Uncoupled	Coupled
Sharma and Aubeny [34]	2011				√	√						√			√	√	
Cao [35]	2010			√													√
Cardoso and Silveira [36]	2010						√						√			√	
Hodder et al. [37]	2010						√					√				√	
Jin et al. [38]	2010						√						√			√	
Kimiaei et al. [39]	2010				√									√		√	
Nakhaee [40, 41]	2008 and 2010									√	√	√		√		√	
Aubeny et al. [18, 42]	2006 and 2009									√	√	√				√	
Randolph and Quiggin [19]	2009									√	√	√			√	√	
Oliphant et al. [43]	2009						√						√			√	

Author	Year	SCR-seabed interaction models					Pipe-seabed interaction models					Cyclic loading		Wave loading		Analysis	
			Analytical models					Analytical models									
		Model tests	γ Rigid seabed	β Linear seabed	α Non-linear seabed	α Trenching effects	Model tests	γ Rigid seabed	β Linear seabed	α Non-linear seabed	α Trenching effects	α Vertical	α Lateral	β Regular	α Irregular	β Uncoupled	α Coupled
Bruton et al. [6, 30]	2006 and 2008						√						√			√	
Clukey et al. [44]	2008						√					√				√	
Palmer [11]	2008		√													√	
Sen [45, 46]	2008			√													√
Xia et al. [47]	2008			√											√		√
Akpan et al. [48]	2007		√														√
Karunakaran et al. [9]	2005			√											√		√
Bridge et al. [17]	2004											√					
Giertsen et al. [49]	2004						√		√	√	√	√					
Bridge et al. [3]	2003	√					√					√					√
Langner [50]	2003		√												√		√
Bridge and Willis [51]	2002	√										√					√
Thethi and Moros [10]	2001	√										√					√

Author	Year	SCR-seabed interaction models					Pipe-seabed interaction models					Cyclic loading		Wave loading		Analysis	
			Analytical models					Analytical models									
			γ	β	α	α		γ	β	α	α	α	α	β	α	β	α
		Model tests	Rigid seabed	Linear seabed	Non-linear seabed	Trenching effects	Model tests	Rigid seabed	Linear seabed	Non-linear seabed	Trenching effects	Vertical	Lateral	Regular	Irregular	Uncoupled	Coupled
Willis and West [52]	2001	√										√					√
Pesce et al. [53]	1998			√													
Verley and Lund [27]	1995						√						√				
Hale et al. [54]	1992						√					√	√				
Dunlap et al. [15]	1990						√					√					
Brennodden et al. [20]	1989						√						√				
Morris et al. [22]	1988						√						√				
Wagner et al. [23]	1987						√						√				

SCR-seabed interaction models: The majority of experimental studies carried out in recent years have presented the non-linear behaviour of SCR-seabed interaction and trenching effects in the TDZ. An SCR in the TDZ was recently identified as a fatigue hotspot that substantially increased fatigue damage under the SCR-seabed interaction phenomenon. Better understanding of the significance of SCR-seabed interaction behaviour and soil properties improves the fatigue life prediction in the TDZ. Most of the existing riser models unrealistically simplified SCR-seabed interaction behaviour by assuming a rigid or linearly elastic seabed. Trench formation and trench deepening have also significant influence on SCR-seabed response.

Pipe-seabed interaction models: Experimental model tests and analytical models of vertically loaded horizontal pipes in clay sediment provided valuable information for better awareness of SCR-seabed interaction in the TDZ. These experimental and analytical data produce the general load/deflection curve for pipe-seabed interaction and necessary information for validation of V-z model and determination of geotechnical parameters used in the model.

Cyclic loading: SCR-seabed interaction processes should cover vertical and lateral responses to cyclic loading. Movement and oscillation of the SCR in the TDZ will cause trenching and dynamic embedment of the SCR into the seabed. Seabed stiffness degradation due to cyclic oscillations has a significant influence on the behaviour of an SCR in the TDZ, and especially on the SCR's strength and fatigue performance. After the seabed soil approaches the maximum strength throughout the applied cyclic loading, the seabed soil tends to lose strength and stiffness with the increase in plastic embedment during cyclic oscillations. The seabed soil stiffness degradation mechanism comprises stiffness reduction presented in uplift, suction, and separation as well as the re-penetration process. The degradation of soil stiffness with cyclic loading is best captured by the non-linear seabed model.

Wave loading: The use of regular wave theories does not adequately represent wave loading on SCRs. However, when used, the level of sophistication in random wave loading is highly variable. For example, the length of simulation used to estimate response levels differs widely. Most studies use simplifying assumptions due to the extensive computational time needed to perform random time domain simulation properly.

Analysis (coupled vs. uncoupled): SCRs have a relative effect on the motions of a floating unit, which in turn affects the SCR fatigue life. In a coupled analysis, the floating unit and SCR are modelled together including their mass, stiffness and damping. Coupled analysis is computationally demanding, especially for robust finite element mesh size. In uncoupled analysis, platform wave frequency is computed in separate models by different programs. Once the floating unit motions are obtained, either from coupled or uncoupled analysis, they are imported as input into the riser analysis software for the uncoupled riser analysis. In uncoupled analysis, the riser is considered to have no effect on the platform at its top. These effects are usually negligible, and an uncoupled analysis is adequate.

5. MORE INVESTIGATION

The literature evaluation presented in this chapter demonstrates that the mobility of the SCR at the TDP is mostly characterized by lateral, vertical, and cyclic movements. SCRs are currently a major focus of research, especially regarding fatigue and their interaction with the seabed at the TDP. The current analysis of the SCR is conducted using simplified models of pipe-seabed interaction and does not take into account the embedding of the SCR into the seabed or the effects of soil suction in the TDZ. This omission will have an impact on the expected response of the SCR. Prior investigations have demonstrated that the soil suction effect has the ability to enhance the bending stress experienced by SCRs in the TDZ. In a deepwater setting, the prevailing soil condition is primarily soft clay soil with a low undrained shear strength. Field observations have revealed that the presence of a trench is a frequent occurrence caused by the penetration of the SCR pipe into the seabed. Nevertheless, there is a scarcity of published work examining the impact of trenching on the dynamic structural behavior and fatigue performance of the SCR's riser pipe in the TDZ. The loss in seabed stiffness caused by cyclic motion is a crucial factor for estimating the fatigue of the SCR in the TDZ. This component has not been thoroughly examined, and the seafloor is traditionally not adequately represented in the existing SCR fatigue study. Prior research has established that the choice of soil model used in fatigue estimation calculations significantly impacts fatigue damage. The seabed non-linear model, used to simulate the interaction between the SCR and the seabed, has been demonstrated to be more advanced than models that use linear soil springs for SCR-seabed interactions.

It can be concluded from the summary of models presented in the existing literature survey that:

- Although a linear seabed model is the common modelling for seabed response [9, 47, 55], which is too simplified to capture the nature of SCR-seabed interaction, SCR-seabed interaction is a considerably non-linear phenomenon. For better understanding of SCR behaviour and reliable prediction of the fatigue life in the TDZ, a numerical study and analysis of SCRs with vessel motions should be performed;
- Fatigue performance assessment of the SCR in the TDZ remains a serious design challenge for SCR behaviour. Despite some research papers presenting a reduction in fatigue damage [34, 40, 50] due to riser embedment in the TDZ, other studies have proposed an increase in fatigue damage [7, 49]. These confounding results due to different geotechnical parameters have been imposed with trenches. The trench deepening, gradual embedment of the riser and soil stiffness have an important influence on the SCR's fatigue life in the TDZ. Furthermore, the soil parameters used in riser-seabed analysis can have a significant effect on fatigue life of SCRs. Therefore, SCR's fatigue damage in the TDZ is a critical design aspect where geotechnical consideration becomes important; and

- Although lateral movements of the SCR can influence the riser performance, as suggested by [42, 49], the adduced SCR-seabed interaction analytical models regard only the vertical SCR motions and neglect lateral soil friction, as presented by [40, 41].

The research gaps stated above are currently being investigated and addressed by the authors.

6. CONCISE SUMMARY

This resource provides the essential foundation on the geotechnical interaction model for SCR issue, enabling interested engineers and researchers to delve into the relevant research literature. The primary aim of this chapter was to examine the different modelling methods employed in recent research to gain a clearer understanding of how pipes and soil interact under cyclic movements. The analysis of the seabed's reaction to riser loads and the formation of trenches is crucial for the development of safe and cost-effective riser designs. Ongoing research on selective catalytic reduction (SCR) technology is primarily aimed at gaining a more comprehensive knowledge of the thermodynamic behavior of the TDZ (thermocline disintegration zone) and its interaction with the soil on the seabed. The interaction between the soil and the riser is a complex process that includes non-linear behavior of the soil, variations in trench width and depth, and the softening of the seabed soil due to repeated pressure. The modelling of the interaction between the seabed and the riser enables the identification and investigation of the impact of physical phenomena, such as lateral resistance, soil suction forces, and vertical seabed stiffness, on the operation of the SCR. The interaction between the non-linear seabed and riser models will define how the seabed response model affects the fatigue of the SCR. A slight alteration in the rigidity of the seabed can lead to a minor adjustment in the amount of stress experienced by a structure, yet this seemingly modest variation can have a substantial impact on the overall durability of the structure. Hence, it is clear that there is a strong requirement for the seabed-riser interaction modelling to be highly realistic. An exhaustive examination of the recent research on the interaction between SCR and the seafloor was presented. Upon examining various literature pertaining to this subject, certain essential knowledge for both current and future research has been obtained, while also identifying existing gaps in the field. This chapter has provided a comprehensive overview of the latest Subsea Cable Repair (SCR) systems that involve interaction with the seabed, as well as the associated analysis techniques. The text has also examined the current theories for modeling the interaction between SCR (Steel Catenary Riser) and the seabed. It provides a thorough explanation of the methodologies currently employed to assess the structural performance of SCR in the TDZ (Touch Down Zone). The author is now investigating and conducting continuing research on the research gap discussed in this chapter.

AT A GLANCE

Most of the time, the landing zone (TDZ) is where cyclic bending stresses are highest. This makes it a key area for fatigue. A lot of study is being done on catenary steel compliant pipelines or risers (SCRs), especially on how long they last after being fatigued, which is strongly affected by the seabed soil conditions in the TDZ. This part looks at some new papers that may have an effect on the interaction between the SCR and the seabed. The first part of the review looks at how the SCR is usually set up. The next part of the paper is a review of recent study that looked at how deepwater SCRs interact with the seabed. The review also talked about the SCR's analysis methods, such as the modeling theory and models for geotechnical response. It is clear what the study gap is and what more research needs to be done.

REFERENCES

1. Mekha BB. New frontiers in the design of steel catenary risers for floating production systems. Journal of Offshore Mechanics and Arctic Engineering. 2001;123:153

2. Sparks C. Fundamentals of Marine Riser Mechanics: Basic Principles & Simplified Analysis. PennWell; 2008

3. Bridge C, Howells H, Toy N, Parke G, Woods R. Full scale model tests of a steel catenary riser. In: Proceding of International Conference on fluid Structure Interaction; Cadiz, Spain. 2003
4. Hatton S. Update on the design of steel catenary riser systems. In: Society of Underwater Technology; Newcastle, UK. 1999
5. Patel M, Seyed F. Review of flexible riser modelling and analysis techniques. Engineering Structures. 1995;17:293-304
6. Bruton DAS, White DJ, Carr M, Cheuk JCY. Pipe-soil interaction during lateral buckling and pipeline walking—The safebuck JIP. In: Offshore Technology Conference; 5-8 May 2008; OTC 19589, Houston, Texas
7. Leira BJ, Passano E, Karunakaran D, Farnes KA, Giertsen E. Analysis guidelines and application of a riser-soil interaction model including trench effects. In: 23rd International Conference on Offshore Mechanics and Arctic Engineering, OMAE2004-51527. Vancouver, British Columbia, Canada: ASME; 2004
8. Karunakaran D, Lund KM, Nordsve NT. Steel catenary riser configurations for north sea field developments. In: Offshore Technology Conference; OTC 10979, Houston, Texas. 1999
9. Karunakaran D, Meling T, Kristoffersen S, Lund K. Weight-optimized SCRs for deepwater harsh environments. In: Offshore Technology Conference; OTC 17224; 2-5 May 2005; Houston, TX, USA
10. Thethi R, Moros T. Soil interaction effects on simple catenary riser response. In: Proceeding Conference on Deepwater Pipeline & Riser Technology; Houston, Texas. 2001
11. Palmer A. Touchdown indentation of the seabed. Applied Ocean Research. 2008:235-238
12. Hetényi M, Michigan U o. Beams on Elastic Foundation. The Univ. of Michigan Press; 1946
13. Jones G. Analysis of Beams on Elastic Foundations: Using Finite Difference Theory. Inst of Civil Engineers Pub; 1997

14. Chen WF, Han DJ. Tubular Members in Offshore Structures. Pitman Publishing Inc.; 1985

15. Dunlap WA, Bhojanala RP, Morris DV. Burial of vertically loaded offshore pipelines in weak sediments. In: Offshore Technology Conference, OTC 6375; 7-10 March 1990; Houston, Texas

16. Barltrop NDP, Adams AJ, Hallam M. Dynamics of Fixed Marine Structures. Oxford: Butterworth-Heinemann; 1991

17. Bridge C, Laver K, Clukey E. Steel catenary riser touchdown point vertical interaction models. In: Offshore Technology Conference, OTC 16628; Houston, Texas. 2004

18. Aubeny C, Biscontin G, Zhang J. Seafloor interaction with steel catenary risers. In: Texas: Offshore Technology Research Center, Texas A&M University College Station, Houston, Final Project Report to Minerals Management Service, OTRC Library Number 9/06A173. 2006

19. Randolph M, Quiggin P. Non-linear hysteretic seabed model for catenary pipeline contact. In: 28th International Conference on Ocean, Offshore and Arctic Engineering, OMAE2009-79259. Honolulu, Hawaii, USA: Proceedings of the ASME 2009; 2009

20. Brennodden H, Lieng J, Sotberg T, Verley R. An energy-based pipe-soil interaction model. In: Offshore Technology Conference; Houston, Texas, USA. 1989

21. Cheuk C, Bolton M. A technique for modelling the lateral stability of on-bottom pipelines in a small drum centrifuge. In: Physical Modelling in Geotechnics—6th ICPMG; London. 2006.

22. Morris DV, Webb RE, Dunlap WA. Self-burial of laterally loaded offshore pipelines in weak sediments. In: Offshore Technology Conference; OTC 5855, Huston, Texas. 1988

23. Wagner D, Murff J, Brennodden H, Sveggen O. Pipe-soil interaction model. In: Offshore Technology Conference; 27-30 April 1987; Houston, Texas, USA. OTC

24. White DJ, Cheuk CY. Modelling the soil resistance on seabed pipelines during large cycles of lateral movement. Marine Structures. 2008;21:59-79

25. Cathie D, Jaeck C, Wintgens J-F Ballard J-C. Pipeline geotechnics—State of the art. In: Proc. International Symposium on Frontiers in Offshore Geotechnics (ISFOG); Perth, Australia. 2005

26. Dendani H, Jaeck C. Flowline and riser: Soil interaction in plastic clays. In: Offshore Technology Conference, OTC 19261; 5-8 May 2008; Houston, Texas, USA

27. Verley R, Lund KM. A soil resistance model for pipelines placed on clay soils. In: American Society of Mechanical Engineers. 1995. pp. 225-225

28. Zhang J, Stewart D, Randolph M. Modeling of shallowly embedded offshore pipelines in calcareous sand. Journal of Geotechnical and Geoenvironmental Engineering. 2002;128:363

29. Lyons C. Soil resistance to lateral sliding of marine pipelines. In: Offshore Technology Conference; 1973; OTC 1876, Texas, USA

30. Bruton D, White D, Cheuk C, Bolton M, Carr M. Pipe/soil interaction behavior during lateral buckling, including large-amplitude cyclic displacement tests by the safebuck jip. In: Offshore Technology Conference; Houston, USA. 2006

31. Bruton D, Carr M, Crawford M, Poiate E. The safe design of hot on-bottom pipelines with lateral buckling using the design guideline developed by the safebuck joint industry project. In: Deep Offshore Technology Conference; Espirito Santo, Brazil. 2005

32. Oliphant J, Maconochie A. The axial resistance of buried an unburied pipelines. In: Offshore Site Investigation and Geotechnics, Confronting New Challenges and Sharing Knowledge. 2007

33. DNV 2007. Dnv-rp-f113: Recommended practice pipeline subsea repair. Det Norske Veritas

34. Sharma P, Aubeny C. Advances in pipe-soil interaction methodology and application for SCR fatigue design. In: Offshore Technology Conference; Houston, USA. 2011

35. Cao J. Fatigue analysis technique of deepwater steel catenary risers. In: 29th International Conference on Ocean, Offshore and Arctic Engineering, OMAE 2010; 6-11 June 2010. Shanghai, China: ASME

36. Cardoso C, Silveira R. Pipe-soil interaction behavior for pipelines under large displacements on clay soils—A model for lateral residual friction factor. In: Offshore Technology Conference; OTC 20767, Houston, Texas. 2010

37. Hodder MS, Byrne BW. 3d experiments investigating the interaction of a model SCR with the seabed. Applied Ocean Research. 2010

38. Jin J, Audibert JM, Kan WC. Practical design process for flowlines with lateral buckling. In: Proceedings of the ASME 2010 29th International Conference on Ocean, Offshore and Arctic Engineering, OMAE2010-20478. Shanghai, China: ASME; 2010

39. Kimiaei M, Randolph M, Ting I. A parametric study on effects of environmental loadings on fatigue life of steel catenary risers (using a nonlinear cyclic riser-soil interaction model). In: Proceedings of the ASME 2010 29th International Conference on Ocean, Offshore and Arctic Engineering; Shanghai, China. ASME; 2010

40. Nakhaee A, and Zhang J. Effects of the interaction with the seafloor on the fatigue life of a SCR. In: Proceedings of the 18th (2008) International Offshore and Polar Engineering Conference; 2008

41. Nakhaee A, Zhang J. Trenching effects on dynamic behavior of a steel catenary riser. Ocean Engineering. 2010;37:277-288

42. Aubeny CPA, Biscontin G. Seafloor-riser interaction model. International Journal of Geomechanics. 2009;9(3)

43. Oliphant J, Maconochie A, White D, Bolton M. Trench interaction forces during lateral SCR movement in deepwater clays. In: Offshore Technology Conference; Houston, Texas, USA. 2009

44. Clukey E, Young A, Dobias J, Garmon S. Soil response and stiffness laboratory measurements of SCR pipe/soil interaction. In: Offshore Technology Conference, OTC 19303, Houston, Texas. 2008

45. Sen TK. Probability of fatigue failure in steel catenary riser in deep water. Journal of Engineering Mechnics. 2006

46. Sen TK, Brown K. Root fatigue in deep water steel catenary risers—A probabilistic approach for assessment of risk. In: Offshore Technology Conference; 5-8 May 2008; Houston, Texas, USA

47. Xia J, Das PK, Karunakaran D. A parametric design study for a semi/SCR system in northern north sea. Ocean Engineering. 2008;35:1686-1699

48. Akpan U, Koko T, Rushton P, Tavassoli A, Else M. Probabilistic fatigue reliability of large diameter steel catenary risers (SCR) for ultra-deepwater operations. In: Proceedings of the 26th International Conference on Offshore Mechanics and Arctic Engineering. California, USA: ASME; 2007

49. Giertsen E, Verley R, Schroder K. Carisima: A catenary riser/soil interaction model for global riser analysis. In: 23rd International Conference on Offshore Mechanics and Arctic Engineering, OMAE2004-51345; Vancouver, British Columbia, Canada. 2004

50. Langner CG. Fatigue life improvment of steel catenary risers due to self-trenching at the touchdown point. In: Offshore Technology Conference, OTC 15104, Houston, Texas, USA; 2003

51. Bridge C, Willis N. Steel catenary risers—Results and conclusions from large scale simulations of seabed interactions. In: International Conference on Deep Offshore Technology. New Orleans, USA: PennWell; 2002

52. Willis NRT, and West PTJ. Interaction between deepwater catenary risers and a soft seabed: Large scale sea trials. In: Offshore Technology Conference; Houston, Texas OTC 13113. 2001

53. Pesce CP, Aranha JAP, Martins CA. The soil rigidity effect in the touchdown boundary-layer of a catenary riser: Static problem.

In: Proceedings of the 8th International Offshore and Polar Engineering Conference; 1998; Monterial, Canada

54. Hale JR, Morris DV, Yen TS, Dunlap WA. Modelling pipeline behaviour on clay soils during storms. In: Offshore Technology Conference, OTC 7019; May 4-7 1992; Houston, Texas

55. Grealish F, Kavanagh K, Connaire A, Batty P. Advanced nonlinear analysis methodologies for SCRs. In: Offshore Technology Conferecne, OTC 18922, Houston, Texas, USA. 2007

04 INVESTIGATION OF LOCAL SCOUR AROUND A MONOPILE FOUNDATION FOR OFFSHORE WIND TURBINES AND ITS IMPACT ON STRUCTURAL RESPONSES

1. INTRODUCTION: AN OVERVIEW

Offshore wind power is becoming a significant global energy source because of its plentiful reserves, high-energy concentration, and reduced civil objections compared to onshore wind power. By the end of 2016, the worldwide installed capacity had surpassed 14,300 MW [1], and there were predictions of even greater potential for expansion in the future [2, 3]. In order to maximize energy harvesting efficiency, the industry is currently progressing towards deeper waters and increasing the size of offshore wind turbines (OWTs) from 5 to 8 MW, 10 MW, and eventually 12 MW [4] (refer to Figure 1). The continuous increase in both sea depth and size of offshore wind turbines (OWTs) has resulted in significant growth in the loading, exacerbating the deformation of the foundations and potentially endangering the operation of the OWTs.

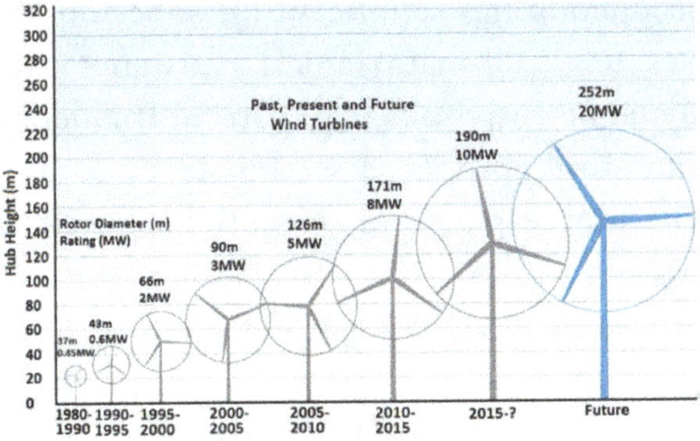

Figure 1. Growth in size and power of horizontal axis wind turbine [4].

As the most commonly adopted foundation for OWTs, the monopile had been used as the foundation of 2653 OWTs by the end of 2016, which accounts for over 80% [5], due to their ease of design and manufacture in contrast to other foundation types. A typical monopile is composed of a hollow steel cylinder with a diameter typically spanning 4 to 8 m and a slenderness ratio (driving depth over diameter) less than 10 [5]. A special aspect to consider with monopile foundations for OWTs is that the vertical load mainly comprised of gravity load is relatively small compared with other offshore structures like oil and gas platforms. The horizontal loads generated by wind and wave actions, which are of similar magnitude to the gravity loads, are therefore of particular importance for the design [6]. A typical laterally loaded offshore monopile foundation generally behaves rigidly due to their large bending stiffness and low slenderness ratio [7]. The serviceability requirements including lateral deformation and stiffness rather than ultimate lateral resistance tend to govern design due to strict rotational tolerance specifications for monopile foundations [8].

Under the current and/or wave loadings, severe local scour can be generated around a monopile situated in cohesionless sediments. The maximum scour depth is generally predicted to be proportional to the pile diameter, and scour depths of over 1.5 times pile diameter were observed in many practical installations [9]. Considering the lower slenderness ratio compared with more conventional offshore pile for oil and gas industries, the maximum scour depth at a monopile may account for over 25% of the driving length [10]. This scour-induced driving length reduction has significant influence on both quasi-static lateral responses (e.g., ultimate lateral resistance and deformation stiffness) and dynamic responses (e.g. natural frequency) of monopile foundations. Therefore, proper assessments of local scour and its influence on structural responses are essential to the design of monopile foundation for OWTs.

In view of the complexity of local scour phenomenon around a monopile and scour effects on manifold monopile responses, this chapter aims to systematically summarize the predicting methods involving the maximum scour depth and the time scale under various flow conditions. The controlling mechanisms of local scour around a monopile are also revisited. Subsequently, we examine existing research on the effects of local scour on the responses of the monopile foundation.

2. LOCAL SCOUR AROUND A MONOPILE

2.1 General description and scour mechanisms

In the river hydraulics, the local scour at the bridge piers has been widely recognized to be one of the main causes for the structure failure [11]. Hence, the phenomenon of scouring at bridge piers has ever been studied extensively over the past half century from various aspects, including local scour mechanisms associated with flow characteristics, prediction of the maximum scour depth, etc. Comprehensive descriptions of scour around a pile or bridge pier in steady currents have been given in Refs. [12, 13].

For a pile foundation installed in the marine environment, several phenomena with regard to the flow pattern would arise including a horseshoe vortex (HSV, or sometimes necklace vortex due to its shape) in front of the pile, wave vortices convected through the pile's lee-side, and the contraction streamlines at the side edges of the pile. These changes induce an increased sediment transport capacity. If the scouring potential created by the large-scale eddy structures is strong enough to overcome the particles' resistance to motion, local scour will be initiated around the pile (see Figure 2). As scour hole develops, the strength of the large-scale eddy structures will gradually be weakened, thereby reducing the transport rate from the local scour region. Equilibrium is reestablished if the sediment inflow of the scour hole is equal to the outflow for live-bed scour, or the shear stress caused by vortices is equal to the critical shear stress of incipient sediment motion at the bottom of the scour hole for clear-water scour. Two major issues concerning the local scouring process, namely the equilibrium scour depth and the time scale, usually need to be quantitatively characterized and are detailed in Sections 2.2 and 2.3, respectively. We herein focus on reexamining the general scouring mechanisms, which mainly involve characterization of the flow pattern under different incoming flow conditions.

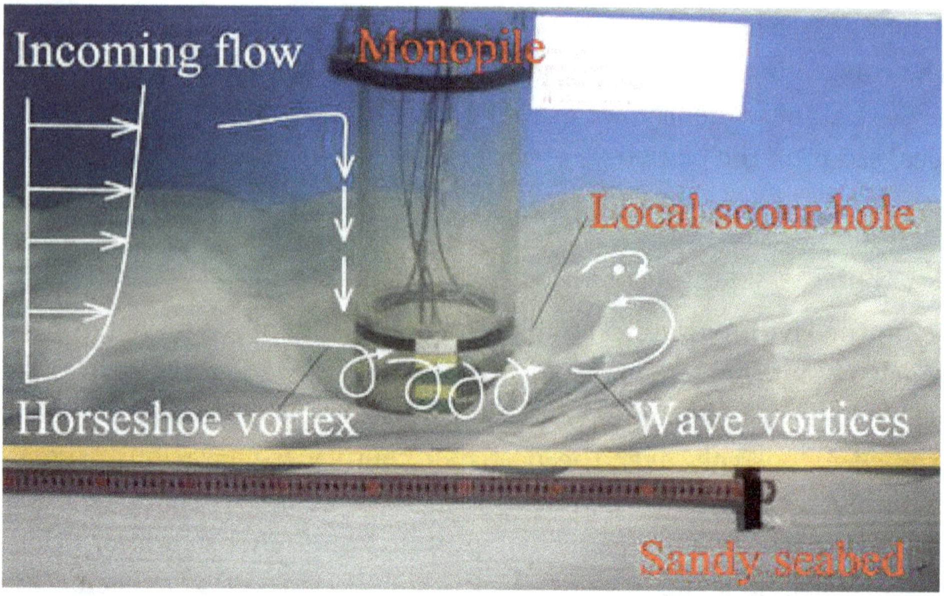

Figure 2. General flow patterns involving local scour around a monopile.

A surface-mounted cylinder induces the boundary layer separation from the upstream bed in the incoming flow, resulting in a HSV system at the upstream junction corner of the cylinder. The existing test observations indicated that the HSV (including the down-flow) upstream of a cylinder is the most evident and significant contributor to the scour process in currents alone. Much work has been done for the HSV at a cylinder in steady currents. Many researchers investigated the HSV at a flatbed for the laminar case [14, 15, 16]. However, most of the practical HSVs occur with turbulent approach boundary layers [17]. Baker [18] used smoke- and oil-flow visualization to observe the turbulent HSV system. The variation in the position of the primary horseshoe vortex with the flow parameters was presented graphically. Dargahi [19] tried to examine the coupling between the flow field and scouring around a circular pile. The consequences of an unsteady vortex pattern were indicated. With the development of the scour hole, the vortex rapidly grows in size and strength as additional fluid attains a downward component and the strength of the down flow increases (see [20]). This conclusion was also verified by the subsequent flow measurements [21, 22].

Kirkil et al. [23] investigated the influence of large-scale coherent structures and macroturbulence events on the scour hole development at a cylinder. The results indicated that the pressure fluctuations and turbulent kinetic energy levels inside the primary necklace vortex were much larger than those associated with the turbulence inside the boundary layer flow approaching the cylinder. Strong ejections and uplift events were observed in the region just behind the cylinder. Recently, the evolution of the HSV in a scour hole around a circular pile was studied by Guan et al. [24] using 2D Particle Image Velocimetry (PIV) technique. The results indicated that the horseshoe vortex system evolves from one initial small vortex to three vortices along the upstream slope of the developing scour hole. Both the strength and size of the main vortex increase with increasing scour hole size. Most existing studies regarding the HSV at a scour hole in steady currents mainly focus on describing the evolution of the HSV itself rather than revealing the time-dependent coupling between the HSV and the scour hole. There is still a great need to improve the understanding of this complex coupling process (see [25]).

In contrast to the substantial amount of accumulated knowledge for HSV in steady currents, relatively fewer studies (e.g., [26]) can be found regarding the flow patterns around a cylinder under waves, although the two-dimensional vortex flow behind a free cylinder subject to an oscillatory flow is well understood (e.g., [27]). Sumer et al. [28] concluded that the scour around a pile in waves is governed by the action of both lee-wake vortices and the horseshoe vortex. Their observations showed that the lee-wake vortices shed from the cylinder act like cyclones, sweeping the sediment grains into its core region and then lifting the grains upward via the updraft in the vortices. The extension and circulation of the lee-wake vortices are governed primarily by the Keulegan Carpenter (KC) number [27]. The KC number is defined as

$$KC = U_{\mathrm{wm}} T / D \tag{1}$$

where U wm is the maximum velocity of the undisturbed wave-induced oscillatory flow at the sea bottom above the wave boundary layer, D is the pile diameter, and T is the wave period. In contrast to the controlling role of the lee-wake vortices in eroding the sediment grains around a cylinder, the HSV would be less significant because the wave boundary layer is usually thin (see [29, 30]), especially for relatively small KC numbers [26]. Regarding monopile foundations for OWTs, the corresponding value of KC number is usually low due to the large diameter [31]. Therefore, the effect of the HSV might be insignificant for the scour development around a monopile under waves.

In general, the controlling mechanism of the scour process around pile foundation in wave-alone or in current-alone condition has been well established. Yet for the condition of combined waves and currents, which is a common scenario in marine environments, the physical mechanism for the scouring phenomenon is less understood. When the waves and currents coexist, the sediment is normally picked up by the waves due to its higher capacity of lifting sands and transported by the currents due to its higher capacity of carrying sands. Nevertheless, the effect of waves and currents' coexistence is more than just a superimposition of their capacities of initiating and carrying sediment, due to a nonlinear interaction between waves and currents within and outside the bottom boundary layer [32, 33, 34]. The only experimental investigation so far on the flow pattern around a vertical pile in a combined current-wave flow was made by Sumer et al. [26]. Their results showed that introducing the following currents in the waves increases the size and lifespan of the horseshoe vortex and lowers the critical KC number for the threshold of horseshoe vortex. The horseshoe vortex exists for smaller KC with increasing the current-to-wave velocity ratio. This result is related to the increase in the adverse pressure gradient in front of the pile caused by the superimposed currents. For small values of KC (such as KC≈O(10) KC≈*O*10), the presence of the horseshoe vortex is quite limited in both space and time domain, and its influence on the scour is much less than that in steady currents (KC≈∞KC≈∞). That is, with decreasing KC, the effect of horseshoe vortex decreases and that of the vortex shedding increases accordingly.

2.2 Equilibrium scour depth

The equilibrium scour depth prediction is a key concern in the geotechnical design for the coastal and offshore foundations. On the basis of laboratory flume tests and field data, many empirical equations have been proposed with respect to the equilibrium scour depth around a cylinder over the past few decades.

For the bridge scour in steady currents, Melville and Sutherland equation [35] and the Colorado State University (CSU) equation [36] are two of the most commonly used scour equations. They give somewhat conservative estimates of scour depth, the former perhaps more so than the latter [37]. Other well-known scour equations were reviewed by Deng and Cai [38]. The accuracy of several most commonly adopted scour equations were evaluated and compared by Johnson [37], Mueller [39] and Sheppard et al. [40], using a large set of field data.

Unlike the pier-scour in bridge waterways, the local scour at offshore monopile foundations should consider the action of waves. For scour depth around pile exposed to waves alone, the number of prediction equation is much fewer compared with that under steady currents. An empirical expression for scour depth at a circular slender pile exposed to regular waves was established [28]:

$$S/D \approx 1.3[1 - \exp(-0.03(KC - 6))] \text{ for } KC \geq 6 \qquad (2)$$

where S is the equilibrium scour depth at the pile. Note that Eq. (2) is valid for the live-bed conditions. Sumer et al. [41] further investigated the effect of the soil density on the scour depth at a pile in waves. The scour depth was significantly increased when the bed soil was changed from medium dense sand to dense sand. Dey et al. [42] presented the results of equilibrium scour depths at circular piles in clay and sand-clay mixed beds under waves alone. The variation of equilibrium scour depth with KC for different soil conditions follows an exponential law, as was given in [28], having different coefficients and exponents. Recently, Zanke et al. [43] proposed the following unifying equation for the prediction of equilibrium scour depth around a pile under the action of waves, tidal or steady currents.

$$S/D = 2.5(1\text{–}0.5\, U_{cr}/U)\, x_{rel} \quad (3)$$

where x rel = x eff /(1 + x eff), x eff = 0.03(1 – 0.35U cr /U) (KC-6), U is the mean velocity in case of steady currents, and U cr is the critical velocity for initiation of sediment motion.

For shallow-water subsea locations, waves often coexist with currents, propagating either following or opposing the currents. This brings the soil response around marine structures more complicated than that due to waves or currents alone. Nevertheless, the related research is relatively few. Sumer and Fredsøe [44] conducted a series of tests of irregular waves propagating either along or perpendicular to the current, indicating that the scour depth for combined waves and currents is influenced by the KC number and the current-to-wave velocity ratio. The current-to-wave velocity ratio is defined as U cw = U c/(U c + U wm), in which U c represents the undisturbed near-bed current velocity component of the combined flow, and U wm is the maximum velocity of the undisturbed wave-induced oscillatory flow at the sea bottom above the wave boundary layer. Sumer and Fredsøe [45] carried out re-analyses of scour data in [44] and derived the following empirical expression for the scour depth under combined waves and current:

$$S/D = S_c/D\{1 - \exp\left[-A(KC - B)\right]\} \text{ for } KC \geq B \quad (4)$$

in which S c is the scour depth for the current-only case, and the parameters A and B are given as $A = 0.03 + 0.75U_{cw}^{2.6}$ and B = 6exp(–4.7U cw). The validity range of Eq. (4) is limited to 4 < KC < 25. The value of Sc /D can be expressed with a mean value Sc /D = 1.3 and a standard deviation S/D = 0.7. The maximum scour depth for the live-bed scour regime is suggested to take as Sc /D = 1.3 + S/D = 2.0 or Sc /D = 1.3 + 2 S/D = 2.7 for design purpose. The large variation range of Sc/D in Eq. (4) could produce a great arbitrariness and uncertainty in the practical engineering application. Moreover, the accuracy of Eq. (4) is not very much optimistic [46]. Rudolph and Bos [47] carried out model tests on scour around a monopile under combined waves and currents with oblique directions, focusing on the range 1 < KC < 10. They proposed an improved scour depth prediction formula based on Sumer and Fredsøe [45] by analyzing their new data and preceding data.

Recently, Qi and Gao [31, 48] conducted an experimental investigation of the local scour development under combined waves and currents at a monopile. The influence of the wave propagating direction relative to currents is obvious. The experimental results indicated that the wave-current interaction had a substantial effect on the time-development of local scour and the resulting equilibrium scour depth. An average-velocity-based Froude number (Fr a) was proposed to correlate with the equilibrium scour depth (S/D) at offshore monopile foundation in combined waves and currents [48]. The definition of Fr a is

$$Fr_{\mathrm{a}} = U_{\mathrm{a}}/\sqrt{gD} \tag{5}$$

in which $U_{\mathrm{a}}\ (=\frac{1}{T/4}\int_0^{T/4}(U_{\mathrm{c}} + U_{\mathrm{wm}}\sin(2\pi t/T))dt = U_{\mathrm{c}} + \frac{2}{\pi}U_{\mathrm{wm}})$ is the average water particle velocity during one-quarter cycle of oscillation under combined waves and currents, when the oscillatory motion and the currents are in the same direction; t is the time and g denotes the gravitational acceleration. The equilibrium scour depths (live-bed scour regime) under combined waves and currents obtained from the experiments were plotted as a function of Fr a (see Figure 3a), along with those of in [44, 49]. Following empirical expression for the correlation between S/D and Fr a was given for predicting equilibrium scour depth under live-bed conditions.

Eq. (6) indicates that the scour depth approaches its mathematical asymptotic value (S/D ≈ 2.0) as Fr a increases (e.g., larger than 1.0). Note that this asymptotic value is in the range of a typical scour depth prediction (S/D ≈ 1.7–2.4) by previous local scour equations around a pile for current-alone conditions. The prediction using the empirical correlation generally distributes in the range of 25% error lines (see Figure 3b), which is superior to the existing formula.

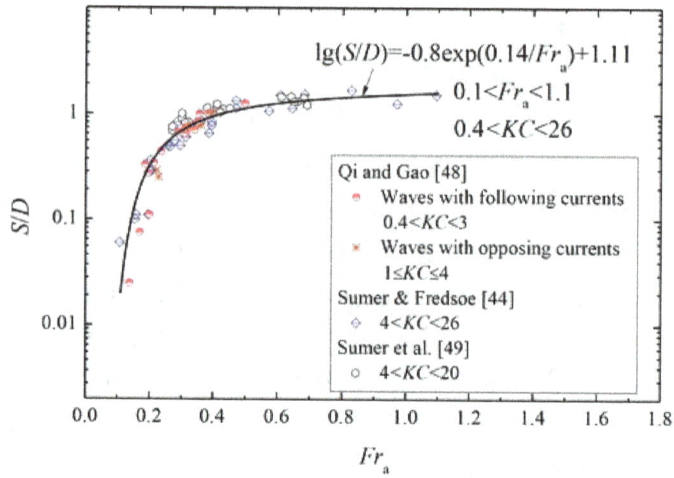

(a)

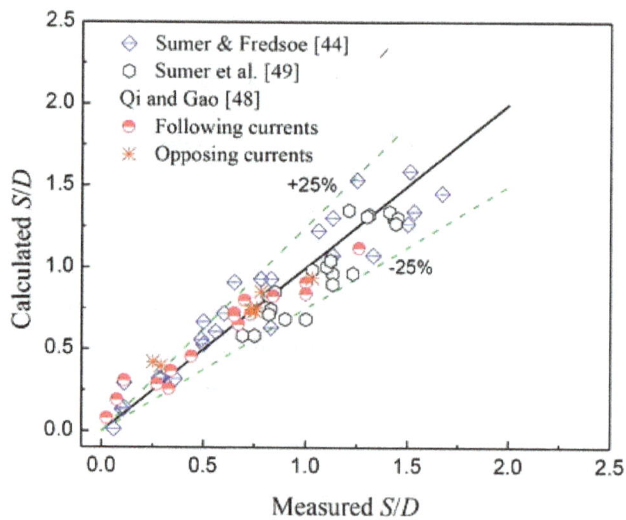

(b)

Figure 3. (a) S/D vs. Fra for combined waves and currents; (b) comparison of measured and predicted scour depths [48].

$$\lg(S/D) = -0.8\exp(0.14/Fr_a) + 1.11 \; (0.1 < Fr_a < 1.1, \; 0.4 < KC < 26) \qquad (6)$$

As aforementioned, most existing equations for predicting the equilibrium scour depth around a cylinder are based predominantly on scaled laboratory flume data. Therefore, the inherent scaling issue is worthy of clarifying before extrapolating the results of small scale scour experiments to prototype [50]. The scale effect is essentially attributable to the use of three independent length scales in local scour experiments: cylinder diameter, sediment size, and water depth [51]. The three length scales lead to several controlling similarity parameters that inevitably cannot be satisfied concurrently. Two aspects are specifically drawing attention involving the scale effect: one associated with inadequate similitude of large scale turbulence generated around a cylinder during scour tests and the other with the distortion of sediment size scaling.

The similitude of large scale turbulence requires preservation of flow patterns so that pressure heads along flow path scale directly with the geometric scale relating the small cylinder in a flume test to a prototype pile. The Froude number defined with incoming flow velocity and pile diameter is proved useful for reflecting the similitude associated with flow-field [51, 52]. It was concluded that the flume-based equations for predicting equilibrium scour depth at a cylinder may not adequately account for the similitude of large scale turbulence and thus commonly lead to over-conservative estimations in prototype applications [51].

Most flume experiments were forced to use particles of comparable size to full-scale particles due to requirements of both the similitude associated with particle mobility and the natural size limit to which bed particles become cohesive. As such, the similarity of sediment size with respect to pile diameter cannot be maintained. According to Melville and Chiew [53], D/d_{50} barely has effect on the scour development for $D/d_{50} > 50$, with d_{50} as the median grain size of the sediment. However, some more recent research indicates a decrease in S/D at very large values of D/d_{50} [54, 55]. Consequently, a prototype scour depth prediction based on the flume scour data is greater than that in the field, that is, a relatively conservative or safe prediction. A comparison between the maximum calculated scour data from different formulations used in standard and technical recommendations for the design of offshore wind farms and available scour data in European wind farms found in technical reports was once conducted by Matutano et al. [9]. The observed maximum scour depths were largely less than estimated in most offshore wind farms studied, which is consistent with the similarity analysis.

It should be noted that various offshore standards like DNV·GL [8] recommend Eq. (2) proposed in [26] to estimate the scour depth around monopile foundations for the OWTs under combined waves and currents. This could be inappropriate and potentially lead to great inaccuracies [56].

2.3 Time scale

The time scale of scour process (T 0) is a certain amount of elapsed time for a substantial amount of scour to develop [45], which is essential for predicting the scour depth at any instant in a certain wave and/or current condition. In the case of live-bed scour, the equilibrium scour depth is achieved faster than the clear-water case, generating a smaller time scale. The time scale of scour process can be defined in several ways according to different mathematical functions for approximating the scour development toward the equilibrium stage [57]. For example, Briaud et al. [58] proposed a hyperbolic function $S_t = S_e\left(\frac{t}{t+T_0}\right)$, where S e is the equilibrium scour depth. Alternatively, a more commonly used definition was proposed in [59]:

$S_t = S_e\left(1 - \exp\left(-\frac{t}{T_0}\right)\right)$. The value of the time scale can be obtained by fitting the measured scour depth development curve with these functions of StSt . The normalized time scale is usually defined as

$$T^* = T_0 \frac{\left(g(s-1)d_{50}^3\right)^{\frac{1}{2}}}{D^2} \tag{7}$$

where s is the specific gravity of sand grains.

For steady currents, the normalized time scale is associated with the ratio of boundary layer thickness to pile diameter ($/D$) and the Shields parameter (θθ), using following empirical expression for live-bed scour regime [59].

$$T^* = \frac{1}{2000}\frac{\delta}{D}\theta^{-2.2} \tag{8}$$

The Shields parameter (θ) is defined as

$$\theta = \frac{U_f^2}{g(\rho_s/\rho_w - 1)d_{50}} \tag{9}$$

in which U f is the maximum value of the undisturbed friction velocity, ρsρs is the sediment grain density, and ρwρw is the water density. The detailed calculation procedure of θ is given by Soulsby [60].

In the case of waves, the time scale of the scour process around a cylinder is governed by two parameters, namely the KC number and the Shields parameter. The empirical expression of T* based on KCKC and θθ for live-bed scour regime can be represented by [59].

$$T^* = 10^{-6}\left(\frac{KC}{\theta}\right)^3 \quad (10)$$

Under combined waves and currents, Petersen et al. [61] concluded that the time scale of scour varies with the KC number, the Shields parameter associated with the wave component of the flow (θwθw), and U cw. The time scale of scour increases significantly when even a slight current is superimposing on a wave. On the basis of the data of Refs. [61, 62] and their own, Chen and Li [63] proposed a piecewise fitting formula for the time scale of scour under combined waves and currents (see Figure 4)

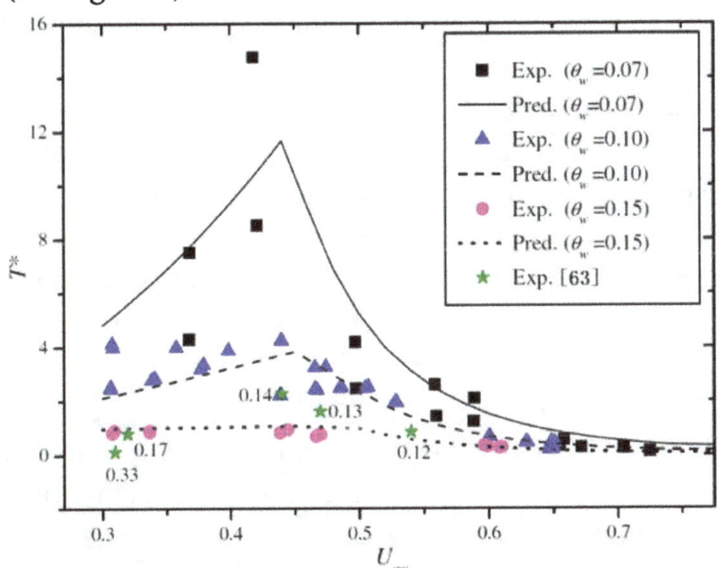

Figure 4. Comparison of predictions using Eq. (11) and calculated results of normalized time scale based on observed data under combined wave-current conditions [63].

$$T_0{}^* = \begin{cases} 8.67 \times 10^{-5}\theta_w^{-5.15}U_{cw}^{28.1(0.152-\theta)} & \text{for } 0.3 < U_{cw} < 0.44 \\ 1.35 \times 10^{-4}\theta_w^{-2.2}U_{cw}^{-6.8} & \text{for } 0.44 \leq U_{cw} \leq 0.7 \end{cases} \quad (11)$$

2.4 Backfilling process

Backfilling process is the deposition of sediment into a scour hole previously generated, when the pile is exposed to a changing flow environment (the flow environment changing from, e.g., a steady current to a wave, from a steady current to combined waves and currents, or from a wave to a smaller wave). For the monopile foundations of the OWTs, the flow climate changes continuously, hence leading to scour and backfilling in an alternating fashion. The backfilling process needs to be quantitatively estimated to establish models for predicting the time history of scour and backfilling for large periods of time (see [64]).

Hartvig et al. [65] experimentally studied the scour and backfilling around a pile subjected to alternating sequences of currents with and without superimposed irregular waves. The results indicated that initially current-scoured beds, when subjected to waves, are backfilled about 10 times slower than the scoured ones. Sumer et al. [49] later presented a systematic investigation of the scour holes backfilling under waves or combined waves and currents around a pile. The results showed that the scour depth corresponding to the equilibrium state of backfilling is the same as that corresponding to the equilibrium state of scour around the pile for the same wave or combined waves and currents climate. That is, the final equilibrium scour is uninfluenced by the loading history of the flow but only dependent on the flow condition at the last stage if the last loading stage is sufficiently long. The time scale of backfilling is much larger than that of scour when the KC number associated with the backfilling is KC < O(10) (typical wind farm application). Various empirical equations involving the time scale were further proposed for different flow environment changes.

Backfilling can reduce the scale of the scour hole, which is beneficial to the bearing behavior of a monopile unprotected against scour. Sørensen et al. [66] assessed the relative density of the backfilled soil around a model monopile of 0.55 m diameter based on the test in a large wave channel. The relative density of the backfill material was found to be rather large, varying between 65 and 80%. Note that these values were obtained from only one set of laboratory tests, and lower densities may exist. The backfilled soil was found to increase the soil support to a monopile foundation significantly. Therefore, an optimized design with respect to both the fatigue limit state and the ultimate limit state can be achieved by considering the time variation of the scour depth [67].

3. EFFECTS OF LOCAL SCOUR ON THE LATERAL RESPONSE OF A MONOPILE

3.1 Lateral bearing behavior

The lateral response of a pile is commonly evaluated using a load transfer approach, with "p-y" curves quantifying the nonlinear interaction between the pile and the surrounding soil. The p-y curves idealize the soil as a series of independent springs distributed along the pile length, with each spring describing the nonlinear relationship between the lateral soil resistance (p) and the lateral deflection of the pile (y). Various formulations for p-y curves have been developed for sands, for example Refs. [68, 69, 70, 71]. The results in Refs. [68, 69] form the basis of the widely used API method [72].

Scour around submarine structures leads to a reduction in the effective stresses and hence strength and stiffness of the soil at a given depth around a pile. As scour depth is predicted to scale with the pile diameter, scouring can impact greatly on the overall response of large-diameter monopiles due to the low slenderness ratio (typically between about 3 and 10). As shown in Figure 5, scour reduces the pile embedment depth from L to L' and increases the load eccentricity (cantilever length) from e to e', changing the profiles of lateral soil resistance p, inducing much larger deformations of the pile and reducing the foundations stiffness.

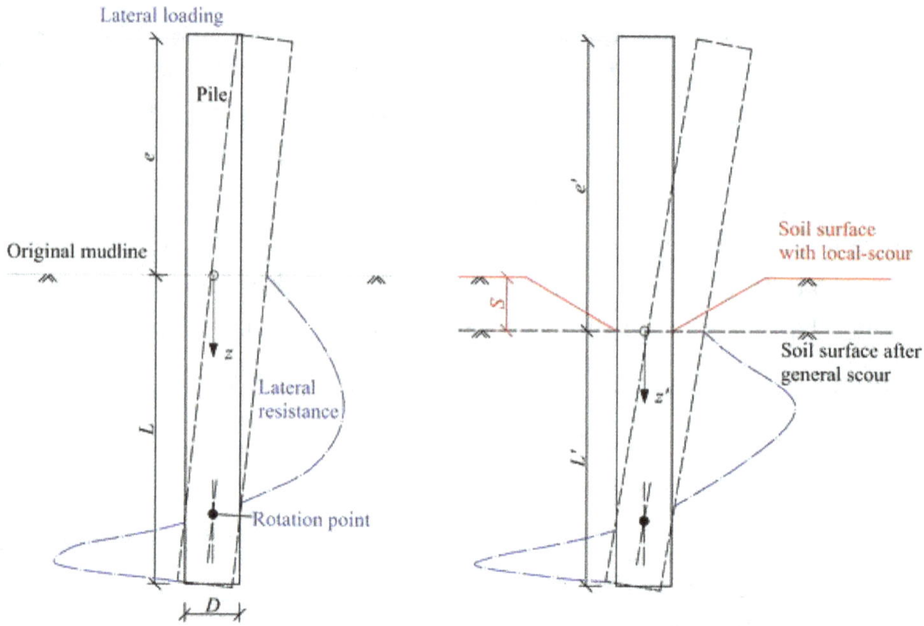

Figure 5. Illustration of scour effect on the lateral response of a shallowly embedded pile [73].

Scour can cause the p-y response at a given absolute depth z (measured from the original soil surface or "mudline") to soften, reducing both initial stiffness and the limiting resistance. In design codes such as DNV·GL [8], local scour is often considered to cause complete loss of soil resistance down to the depth of scour below the original seabed while estimating the pile capacities. Although such a design practice is simple, it often leads to extreme conservative pile foundation design and subsequent unnecessary cost [73]. In recent years, the importance of scour hole dimensions for pile capacities has been increasingly recognized.

In an analytical study [74], general scour effects were incorporated by updating the parameters per unit length (p ult) at a given depth to account for the over-consolidation induced by removal of a scour depth of sand. Lin et al. [75] and Yang et al. [76] further modified the ultimate lateral resistance due to changes in the shallow wedge-type failure mechanism to consider the effects of the local scour. Results indicated that the local scour hole would result in much higher lateral soil resistance for a given depth than for the general scour case, that is, complete removal of the soil surface layer. Tseng et al. [77] combined the approach of Sørensen [78] for modifying the initial modulus of subgrade reaction (k py) and the approach of Lin et al. [75] for modifying p ult to evaluate the scour effect on the deformation response of monopile foundations. According to the case analysis, when the p-y curve method suggested by the design guidelines for an offshore wind turbine was used to design the monopile foundation for an OWT, the foundation deformation was underestimated for a scour depth of less than pile diameter, and foundation stiffness was underestimated for a scour depth of greater than pile diameter.

A series of centrifuge tests were conducted by Qi et al. [73] to investigate the scour effects on p-y curves. A practical approach to incorporate effects of scour on the p-y curves was proposed by adopting an effective soil depth (z e) in the determination of p-y curves. The effective soil depth was expressed by

$$\frac{z_e}{D} = \frac{z'}{D} + \tanh\left(f\frac{z'}{D}\right)\frac{S}{D} \qquad \begin{cases} f \approx 1.5 & \text{for local-scour} \\ f = 0 & \text{for general-scour} \end{cases} \tag{12}$$

where the effective soil depth zeze is a weighted-average of the soil depth relative to the original mudline z and the soil depth below the current scour base z′z′ (=z−S=z−S , S is the scour depth), f is an empirical parameter indicating the transition rate for the effective soil depth from z e = z’ at the current mudline to ze≈zze≈z as the soil depth increases. The recommended value of f = 1.5 for local-scour conditions was based on the tests with 30° slope angle of the scour hole. As the slope angle of the scour hole decreases, the effect of the sloping overburden soil above the level of the scour base reduces and the value of f should be reduced accordingly. This conclusion conforms to the results of a recent study by Lin and Wu [79].

Local scour not only exaggerates the deformation and reduces the stiffness of the monopile foundation, but also changes its bearing mechanism. Monopile foundations for the OWTs generally behave rigidly, rotating as a whole and developing a “toe-kick” phenomenon. Figure 6 illustrates the mobilized soil resistance components of a laterally loaded short pile. Significant base shear and axial resistance will be mobilized at the pile tip if the slenderness ratio gets sufficiently small. Moreover, vertical side resistances emerge around the pile shaft due to the mobilized horizontal displacement of the pile tip and rotation of the pile. Numerical investigations by Bekken [80] and Byrne et al. [81] have confirmed the significant contributions due to the horizontal pile tip displacement and rotation of the pile on the resistance of a monopile.

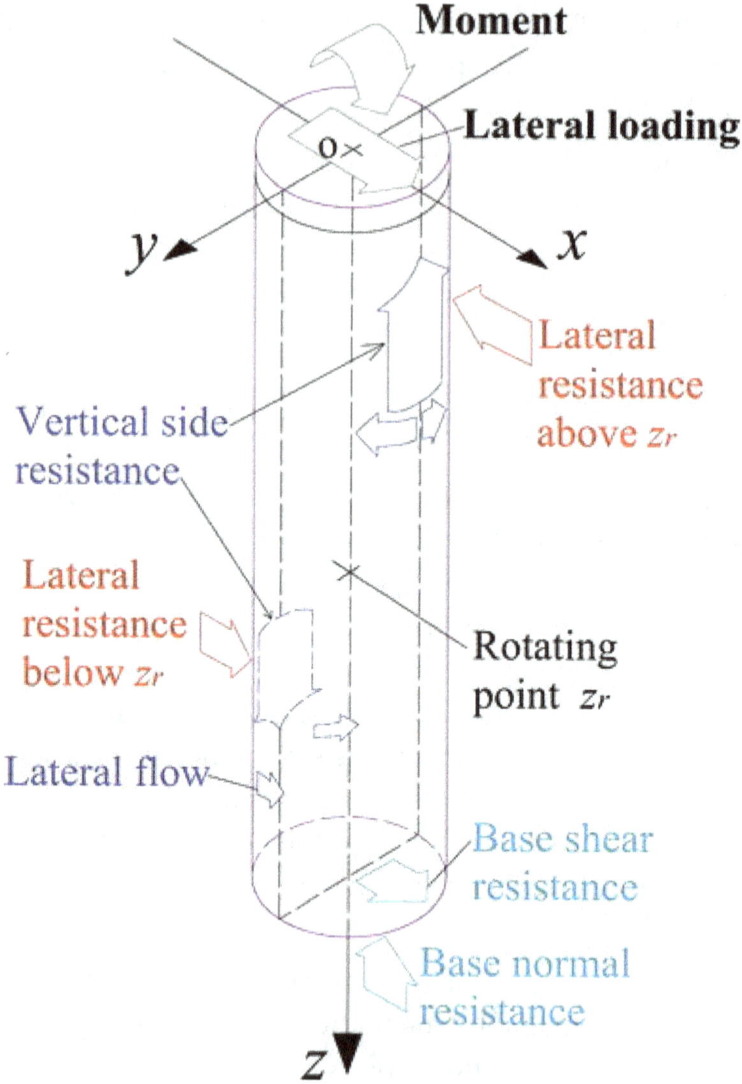

Figure 6. Resistance components of a laterally loaded monopile foundation [82].

To investigate the scour effect on the behavior of a laterally loaded monopile, a three-dimensional finite element model was proposed and verified by Qi and Gao [83]. The results indicated that the scour could induce a significant transition of pile behavior for a typical monopile foundation of the OWTs. Figure 7(a) shows the variation of y_p/y_n with $E_s L^4/E_p I_p$, in which y_n denotes the lateral pile displacement at the loading position obtained from the 3D finite element results, and y_p denotes the lateral pile displacement according to the conventional p-y method. The non-dimensional expression of $E_s L^4/E_p I_p$ is aimed to reflect the rigidity of the pile foundation [7, 84], where I_p denotes the second moment of area of a pile and $E_p I_p$ represents the bending stiffness of the pile. It is seen that for an approximate range of $E_s L^4/E_p I_p > 50$, the value of y_p/y_n is generally equal to 1.0, while the value of y_p/y_n obviously deviates from the value of $y_p/y_n = 1.0$ for $E_s L^4/E_p I_p < 50$. The deviation of y_p/y_n implies that due to the scour-induced increase of pile rigidity (see Figure 7(b)), the pile's tip and shaft resistance could become a significant component of the whole soil resistance. The parameter L_s in Figure 7(b) denotes the pile embedment depth after scour. The traditional p-y approach only takes account of the lateral soil resistance and thus the scour would render the traditional p-y approach inapplicable for post-scour conditions.

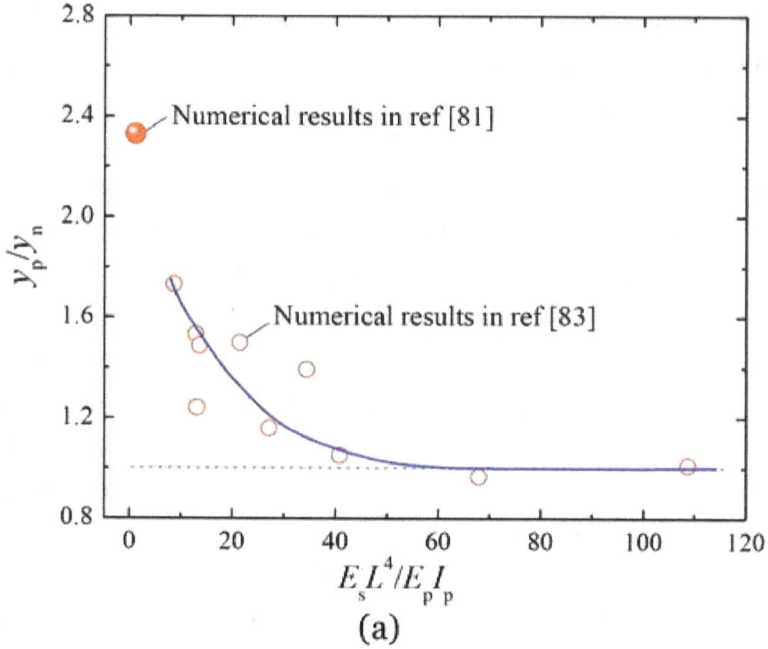

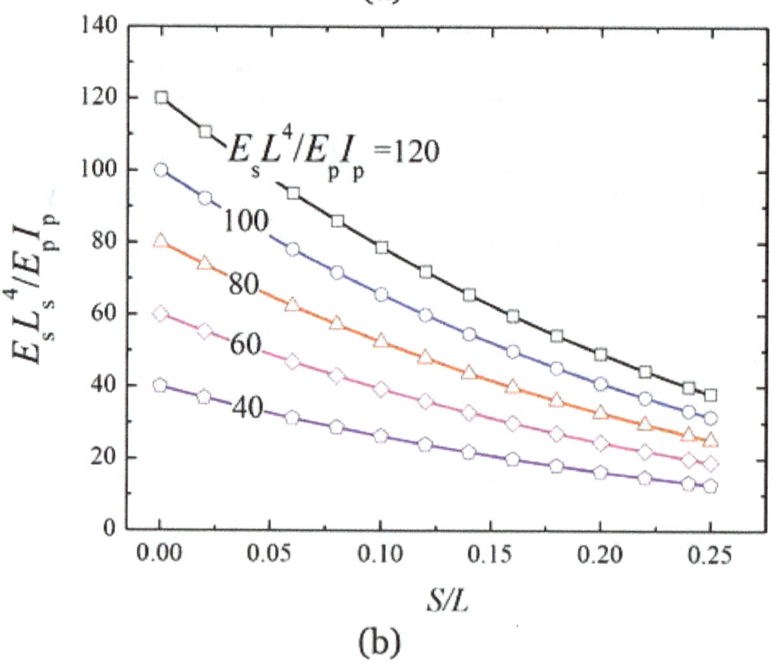

Figure 7. (a) Variation of yp/yn with EsL4/EpIp; (b) effect of scour depth normalized with pile embedment depth S/L on EsLs 4/EpIp [83].

There are relatively fewer studies concerning the scour effect on the pile behavior under cyclic loading compared with aforementioned studies on the monotonic pile behavior [85]. Achmus et al. [86] numerically quantified the scour-induced increase of pile deformation by adopting a degradation stiffness model to account for cyclic loading. The results indicated an evident accumulation of lateral deformation under cyclic loading. Up to now, no systematic investigations into the influence of scour on the cyclic pile behavior can be found in the literatures. This issue needs to be further studied. Moreover, local scour and backfilling occurs around a monopile foundation in an alternating fashion due to the continuously changing flow climate in marine conditions. This variation of scour depth should also be included while evaluating the long-term cyclic pile behavior.

3.2 Natural frequency and fatigue life (FL)

Compared to other types of OWT foundations such as gravity base, tripod, and suction bucket, the monopile is a dynamically sensitive structure due to its flexibility. The primary excitation forces arise from the wind, waves, and the rotor excitations. The fundamental frequency f 0 (the first tower bending frequency) of the structure is the most important controlling parameter for its dynamic response.

The OWTs are usually designed as soft-stiff structures in order to avoid resonances, meaning that f 0 is above the rotational frequency of the rotor f 1P but below the blade-passing frequency f 3P (see Figure 8). The excitation frequencies related to environmental loads from wind and waves are typically below the 1P frequency for OWTs with relatively small installed capacities, for example, smaller than 3.6 MW [87, 88]. Nevertheless, for a 6–8 MW modern OWT on a monopile, designed for the soft-stiff frequency range, f 0 would be around 0.20–0.23 Hz [89]. In this frequency range, the fundamental mode of the structure would resonate with the waves at low wind speeds and be exposed to high spectral energy of waves during higher wind speeds. Due to the narrow soft-stiff frequency range for the OWTs on monopile foundations, even a small natural frequency shift from the initial design value may finally lead to resonance and the resulting stress in the materials of the monopile will be higher than in the quasi-static case. This high stress would potentially reduce the FL of the OWTs significantly.

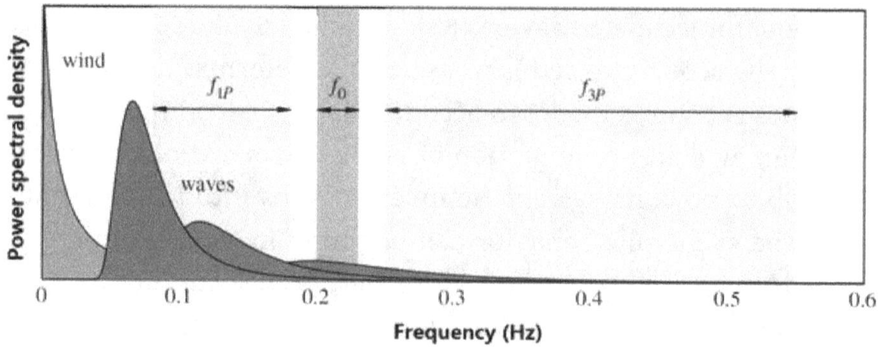

Figure 8. Illustration of typical excitation ranges of a modern 6–8 MW OWT. Wave spectra represent different wind speeds. Not in scale [89].

Scour around the monopile leads to significant reduction of f 0. This would make f 0 align with the frequency of the passing sea waves or the rotational frequency of the rotor (f 1p) under certain circumstances, exposing structures to the threat of resonances. Sørensen and Ibsen [67] calculated the undamped natural frequencies of an existing offshore wind farm at HornsRev1 by means of the Winkler model approach. The stiffness derived from p-y curves given in the API code was adopted in their desk study, without accounting for the degradation of stiffness due to cyclic loading. The results revealed that the first natural frequency decreases with approximately 5% when the scour depth reached 1.3D. Prendergast et al. [90] conducted finite element numerical modeling to examine the scour effects on the natural frequency of a 3.6 MW OWT on a monopile in typical offshore sand deposits. The numerical model was validated by scale model tests and proved capable of tracking the scour-induced change in the natural frequency of the system. A reduction of up to 8.5% in natural frequency was observed in loose sand for S/D = 1.3. A comparison among cases with different soil stiffness profiles suggests that scour-induced resonance of an OWT on a monopile is most likely to occur in loose deposits. Prendergast et al. [91] recently improved the numerical model to further include the effect of spatial variability in soil properties derived from measured Cone Penetration Test (CPT) data. A stochastic ground model was developed to account for the uncertainty in soil-structure interaction stiffness for a given offshore site. The results indicated that the first natural system frequency reduces while the range of predicted system frequencies increases as the scour progresses. Therefore, the accuracy of using frequency-based SHM (structural health monitoring) technique to evaluate the magnitude of scour merely based on first natural frequency data decreases as the scour hole goes deeper. Several recent studies have specifically examined the scour impacts that are linked to the foundation loading of offshore wind turbines (OWTs). Damgaard et al. [92] conducted comprehensive aero-hydro-elastic simulations to assess the impact of scour on the fatigue loads of stationary offshore wind turbines (OWTs). Experimental evidence demonstrates that a scour hole with a depth of 1.3 times the diameter (D) results in a significant 40% increase in the fatigue damage equivalent moment. It is important to consider that the changes in stiffness and damping of the soil over time, caused by repeated loading, as well as the variations in scour depth owing to backfilling, have not been taken into consideration. Rezaei et al. [95] conducted a study on the erosion and refilling impacts on the foundation level of a 5 MW monopile supported offshore wind turbine. They included quasi-static, modal, and time-domain fatigue studies in their research. Experimental evidence showed that a scour depth equivalent to 1.5 times the diameter of the wind turbine might lead to a significant decrease of 45% in the fatigue life of the construction. Backfilling is advantageous for extending the fatigue life of offshore wind turbines. However, the relative density has minimal impact on the FL (flutter limit) of the OWTs (offshore wind turbines).

4. CONCISE SUMMARY

The monopile is the prevailing choice for the foundation of offshore wind turbines (OWTs). The presence of waves and/or currents can significantly compromise the functionality of the structure system due to local scour. Unlike bridge scour, which has been extensively researched for the past 50 years, the scour phenomenon around an offshore monopile is characterized by different flow conditions in a marine setting, including waves only, currents solely, and mixed waves and currents. Considering the intricate nature of the incoming flow circumstances, this paper examines various important factors pertaining to local scour around a monopile foundation. These factors include the underlying mechanisms of scour, the deepest point of scour, the duration of the scour process, and the process of backfilling. The inherent problem of scaling that arises from extrapolating the results of small-scale scour studies to a prototype is made clear. The local scour caused by the low slenderness ratio can greatly decrease the driving depth of an offshore monopile, which in turn has a substantial impact on both structural responses. Initially, the process of scouring will lead to a decrease in the stiffness and resistance of the material at a specific depth, resulting in a softening effect. Furthermore, the occurrence of scour would make the conventional p-y technique unsuitable for analyzing post-scour circumstances because of the alteration in the bearing mechanism. Furthermore, conducting a thorough search around the monopile results in a substantial decrease in the fundamental frequency. Under some conditions, this would cause the fundamental frequency to synchronize with the frequency of the ocean waves or the rotational frequency of the rotor, potentially subjecting buildings to the risk of resonances. The present increasing patterns in both water depth and size of offshore wind turbines (OWTs) have resulted in significant growth mostly in the horizontal loading. Monopiles with a maximum diameter of 10 m have been suggested as a viable choice for supporting offshore wind turbines (OWTs) with an installed capacity of around 10 MW. In order to effectively reduce costs while maintaining safety, it is necessary to carefully evaluate the design of monopiles, including predicting scour depth and evaluating the impact of scour on structural responses. A different hybrid foundation design, consisting of a monopile and a bearing plate (or skirted footing), has been proven to improve the lateral bearing capacity in comparison to a monopile [96, 97, 98, 99]. However, the increased effectiveness of the hybrid foundation mostly depends on the correct interaction between the bearing plate and the soils in the shallow zone of the seabed. This is where local scour and wave-induced pore pressure may occur [100, 101, 102, 103, 104, 105, 106, 107, 108]. Further investigation is required to determine the magnitude of local scour and its impact on the bearing capacity and natural frequency of a hybrid monopile.

AT A GLANCE

The foundation type most frequently employed for offshore wind turbines is the monopile. The local erosion caused by the shear flow impacting the monopile foundation has a substantial impact on both the quasi-static lateral behavior and dynamic responses of the monopile. This chapter specifically examines the characterization of local scour in terms of both geographical and temporal dimensions, as well as the identification of scour mechanisms related to the flow field surrounding a monopile. The methods for forecasting the equilibrium scour depth and the time scale of scour are described in detail for different flow conditions in the marine environment. The significance of the scale effect in translating model test results to prototype circumstances is emphasized. Local scour has a considerable impact on the deformation, stiffness, natural frequency, and fatigue life of the monopile foundation and the structure system. Monopiles with diameters of up to 10 m have become a viable choice since the industry is progressing towards deeper waters. Further thorough analysis is required to optimize monopile design by accurately predicting scour depth and evaluating the impact of scouring, with the aim of minimizing costs without compromising safety.

REFERENCES

1. GWEC. Global Wind Report 2016. 8th National Renewable Energy Forum. Ulaanbaatar, Mongolia: Global Wind Energy Council; 2017

2. Perveen R, Kishor N, Mohanty SR. Off-shore wind farm development: present status and challenges. Renewable and Sustainable Energy Reviews. 2014;29:780-792. DOI: 10.1016/j.rser.2013.08.108

3. Wu XN, Hu Y, Li Y, Yang J, et al. Foundations of offshore wind turbines: A review. Renewable and Sustainable Energy Reviews. 2019;104:379-393. DOI: 10.1016/j.rser.2019.01.012

4. Igwemezie V, Mehmanparast A, Kolios A. Current trend in offshore wind energy sector and material requirements for fatigue resistance improvement in large wind turbine support structures—A review.

Renewable and Sustainable Energy Reviews. 2019;101:181-196. DOI: 10.1016/j.rser.2018.11.002

5. Negro V, López-Gutiérrez JS, Esteban MD, Alberdi P, Imaz M, José-María S. Monopiles in offshore wind: preliminary estimate of main dimensions. Ocean Engineering. 2017;133:253-261. DOI: 10.1016/j.oceaneng.2017.02.011

6. Kuo YS, Achmus M, Abdel-Rahman K. Minimum embedded length of cyclic horizontally loaded monopiles. Journal of Geotechnical and Geoenvironmental Engineering. 2012;138(3):357-363. DOI: 10.1061/(ASCE)GT.1943-5606.0000602

7. LeBlanc C, Houlsby GT, Byrne BW. Response of stiff piles in sand to long-term cyclic lateral loading. Géotechnique. 2010;60(2):79-90. DOI: 10.1680/geot.7.00196

8. DNV·GL. DNV-OS-J101: Design of Offshore Wind Turbine Structures. 2013

9. Matutano C, Negro V, JS L–G, et al. Scour prediction and scour protections in offshore wind farms. Renewable Energy. 2013;57/:358-365. DOI: 10.1016/j.renene.2013.01.048

10. Sørensen SPH, Ibsen LB, Frigaard P. Experimental evaluation of backfill in scour holes around offshore monopoles. In: Frontiers in Offshore Geotechnics II. London, UK: Taylor & Francis Group; 2011. pp. 617-622

11. Melville BW, Coleman SE. Bridge Scour. CO, USA: Water Resources Publications; 2000

12. Hoffmans GJCM, Verheij HJ. Scour Manual. Rotterdam: A.A. Balkema; 1997

13. Whitehouse R. Scour at Marine Structures: A Manual for Practical Applications. London: Thomas Telford; 1998

14. Baker CJ. The laminar horseshoe vortex. Journal of Fluid Mechanics. 1979;95(2):347-367. DOI: 10.1017/S0022112079001506

15. Seal CV, Smith CR, Rockwell D. Dynamics of the vorticity distribution in endwall junctions. AIAA Journal. 1997;35(6):1041-1047. DOI: 10.2514/2.192

16. Bo H, Hua Z, Younis MY, Yan L, Raza MS. Experimental investigation on the transition of separation/attachment in steady laminar juncture flows. Experiments in Fluids. 2015;56(4):1-9. DOI: 10.1007/s00348-015-1943-5

17. Simpson RL. Junction flows. Annual Review of Fluid Mechanics. 2001;33(33):415-443

18. Baker CJ. The turbulent horseshoe vortex. Journal of Wind Engineering and Industrial Aerodynamics. 1980;6(1–2):9-23. DOI: 10.1016/0167-6105(80)90018-5

19. Dargahi B. Controlling mechanism of local scouring. Journal of Hydraulic Engineering. 1990;116(10):1197-1214. DOI: 10.1061/(ASCE)0733-9429(1990)116:10(1197)

20. Breusers H, Nicollett G, Shen H. Local scour around cylindrical piers. Journal of Hydraulic Research. 1977;15:211-252. DOI: 10.1080/00221687709499645

21. Dey S, Raikar RV. Characteristics of horseshoe vortex in developing scour holes at piers. Journal of Hydraulic Engineering. 2007;133(4):399-413.DOI:10.1061/(ASCE)0733-9429(2007)133:4(399)

22. Unger J, Hager WH. Down-flow and horseshoe vortex characteristics of sediment embedded bridge piers. Experiments in Fluids. 2007;42(1):1-19. DOI: 10.1007/s00348-006-0209-7

23. Kirkil G, Constantinescu SG, Ettema R. Coherent structures in the flow field around a circular cylinder with scour hole. Journal of Hydraulic Engineering. 2008;134(5):572-587. DOI: 10.1061/(ASCE)0733-9429(2008)134:5(572)

24. Guan DW, Chiew YM, Wei MX. Characterization of horseshoe vortex in a developing scour hole at a cylindrical bridge pier. International Journal of Sediment Research. 2019;34(2):118-124. DOI: 10.1016/j.ijsrc.2018.07.001

25. Ettema R, Constantinescu G, Melville BW. Flow-field complexity and design estimation of pier-scour depth: Sixty years since Laursen and Toch. Journal of Hydraulic Engineering. 2017;143(9):03117006. DOI: 10.1061/(ASCE)HY.1943-7900.0001330

26. Sumer BM, Christiansen N, Fredsøe J. Horseshoe vortex and vortex shedding around a vertical wall-mounted cylinder exposed to waves. Journal of Fluid Mechanics. 1997;332:41-70. DOI: 10.1017/S0022112096003898

27. Williamson CHK. Sinusoidal flow relative to circular cylinders. Journal of Fluid Mechanics. 1985;155:141-174. DOI: 10.1017/S0022112085001756

28. Sumer BM, Fredsøe J, Christiansen N. Scour around a vertical pile in waves. Journal of Waterway, Port, Coastal, and Ocean Engineering. 1992;118(1):15-31. DOI: 10.1061/(ASCE)0733-950X(1992)118:1(15)

29. Jonsson IG. Wave boundary layers and friction factors. In: Proceedings of the Tenth Conference on Coastal Engineering - American Society of Civil Engineers; 1966. pp. 127-148. DOI: 10.1061/9780872620087.010

30. Baykal C, Sumer BM, Fuhrman DR, Jacobsen NG, Fredsøe J. Numerical simulation of scour and backfilling processes around a circular pile in waves. Coastal Engineering. 2017;122:87-107. DOI: 10.1016/j.coastaleng.2017.01.004

31. Qi WG, Gao FP. Physical modeling of local scour development around a large-diameter monopile in combined waves and current. Coastal Engineering. 2014;83:72-81. DOI: 10.1016/j.coastaleng.2013.10.007

32. Kemp PH, Simons RR. The interaction between waves and a turbulent current: Waves propagating with the current. Journal of Fluid Mechanics. 1982;116:227-250. DOI: 10.1017/S0022112082000445

33. Kemp PH, Simons RR. The interaction between waves and a turbulent current: Waves propagating against the current. Journal of Fluid Mechanics. 1983;130:73-89. DOI: 10.1017/S0022112083000981

34. Olabarrieta M, Medina R, Castanedo S. Effects of wave-current interaction on the current profile. Coastal Engineering. 2010;57:643-655. DOI: 10.1016/j.coastaleng.2010.02.003

35. Melville BW, Sutherland AJ. Design method for local scour at bridge piers. Journal of Hydraulic Engineering. 1988;114(10):1210-1226. DOI: 10.1061/(ASCE)0733-9429(1988)114:10(1210)

36. Richardson EV, Davis SR. Evaluating scour at bridges. In: Hydraulic Engineering Circular No. 18. 4th ed. U.S. Fed. Hwy. Admin., U.S. Dept. of Transp., FHWA–NHI–01–001; 2001

37. Johnson PA. Comparison of pier-scour equations using field data. Journal of Hydraulic Engineering. 1995;121(8):626-629. DOI: 10.1061/(ASCE)0733-9429(1995)121:8(626)

38. Deng L, Cai CS. Bridge scour: Prediction, modeling, monitoring, and countermeasures. Practice Periodical on Structural Design and Construction. 2009;15(2):125-134. DOI: 10.1061/(ASCE)SC.1943-5576.0000041

39. Mueller DS. Local scour at bridge piers in nonuniform sediment under dynamic conditions [PhD thesis]. Colorado: Colorado State University; 1996

40. Sheppard DM, Melville B, Demir H. Evaluation of existing equations for local scour at bridge piers. Journal of Hydraulic Engineering. 2013;140(1):14-23. DOI: 10.1061/(ASCE)HY.1943-7900.0000800

41. Sumer BM, Hatipoglu F, Fredsøe J. Wave scour around a pile in sand, medium dense, and dense silt. Journal of Waterway, Port, Coastal, and Ocean Engineering. 2007;133(1):14-27. DOI: 10.1061/(ASCE)0733-950X(2007)133:1(14)

42. Dey S, Helkjær A, Mutlu Sumer B, et al. Scour at vertical piles in sand-clay mixtures under waves. Journal of Waterway, Port, Coastal, and Ocean Engineering. 2011;137(6):324-331. DOI: 10.1061/(ASCE)WW.1943-5460.0000095

43. Zanke UCE, Hsu TW, Roland A, Oscar L, Reda D. Equilibrium scour depths around piles in noncohesive sediments under currents and waves. Coastal Engineering. 2011;58:986-991. DOI: 10.1016/j.coastaleng.2011.05.011

44. Sumer BM, Fredsøe J. Scour around pile in combined waves and current. Journal of Hydraulic Engineering. 2001;127(5):403-411. DOI: 10.1061/(ASCE)0733-9429(2001)127:5(403)

45. Sumer BM, Fredsøe J. The Mechanics of Scour in the Marine Environment. Singapore: World Scientific; 2002. DOI: 10.1142/4942

46. Rudolph D, Bos KJ, Luijendijk AP, et al. Scour around offshore structures—Analysis of field measurements. In: Proceedings of the 2nd International Conference on Scour and Erosion; Vol. 1; 2004. pp. 14-17

47. Rudolph D, Bos K. Scour around a monopile under combined wave-current conditions and low KC-numbers. In: Proceedings of the Sixth International Conference on Scour and Erosion; 2006. pp. 582-588

48. Qi WG, Gao FP. Equilibrium scour depth at offshore monopile foundation in combined waves and current. Science China Technological Sciences. 2014;57(5):1030-1039. DOI: 10.1007/s11431-014-5538-9

49. Sumer BM, Petersen TU, Locatelli L, et al. Backfilling of a scour hole around a pile in waves and current. Journal of Waterway, Port, Coastal, and Ocean Engineering. 2013;139(1):9-23. DOI: 10.1061/(ASCE)WW.1943-5460.0000161

50. Heller V. Scale effects in physical hydraulic engineering models. Journal of Hydraulic Research. 2011;49(3):293-306. DOI: 10.1080/00221686.2011.578914

51. Ettema R, Kirkil G, Muste M. Similitude of large-scale turbulence in experiments on local scour at cylinders. Journal of Hydraulic Engineering. 2006;132(1):33-40. DOI: 10.1061/(ASCE)0733-9429(2006)132:1(33)

52. Ettema R, Melville BW, Barkdoll B. Scale effect in pier-scour experiments. Journal of Hydraulic Engineering. 1998;124(6):639-642. DOI: 10.1061/(ASCE)0733-9429(1998)124:6(639)

53. Melville BW, Chiew YM. Time scale for local scour at bridge piers. Journal of Hydraulic Engineering. 1999;125(1):59-65. DOI: 10.1061/(ASCE)0733-9429(1999)125:1(59)

54. Sheppard DM, Odeh M, Glasser T. Large scale clear-water local pier scour experiments. Journal of Hydraulic Engineering. 2004;130(10):957-963. DOI: 10.1061/(ASCE)0733-9429(2004)130:10(957)

55. Lee S, Sturm T. Effect of sediment size scaling on physical modeling of bridge pier scour. Journal of Hydraulic Engineering. 2009;135(10):793-802. DOI: 10.1061/(ASCE)HY.1943-7900.0000091

56. Negro V, López-Gutiérrez JS, Esteban MD, Matutano C. Uncertainties in the design of support structures and foundations for offshore wind turbines. Renewable Energy. 2014;63:125-132. DOI: 10.1016/j.renene.2013.08.041

57. Qi WG, Li YX, Xu K, Gao FP. Physical modelling of local scour at twin piles under combined waves and current. Coastal Engineering. 2019;143:63-75. DOI: 10.1016/j.coastaleng.2018.10.009

58. Briaud JL, Ting FCK, Cheng HC, Gudavalli R, Perugu S, Wei GS. SRICOS: Prediction of scour rate in cohesive soils at bridge piers. Journal of Geotechnical and Geoenvironmental Engineering. 1999;125(4):237-246. DOI: 10.1061/(ASCE)1090-0241(1999)125:4(237)

59. Sumer BM, Christiansen N, Fredsoe J. Time scale of scour around a vertical pile. In: Proceedings of the Second International Offshore and Polar Engineering Conference; San Francisco; 1992. pp. 308-315

60. Soulsby R. Dynamics of Marine Sands. UK: Thomas Telford; 1997

61. Petersen TU, Sumer BM, Fredsøe J. Time scale of scour around a pile in combined waves and current. In: International Conference on Coastal Engineering; 2012. pp. 971-988

62. Dyrseth S. Time scales for scour below pipelines and around vertical piles in nonlinear random waves and current [Master thesis]. Trondheim: Norwegian University of Science and Technology; 2015

63. Chen B, Li S. Experimental study of local scour around a vertical cylinder under wave-only and combined wave-current conditions in a large-scale flume. Journal of Hydraulic Engineering. 2018;144(9):04018058. DOI: 10.1061/(ASCE)HY.1943-7900.0001502

64. Harris J, Whitehouse RJS, Benson T. The time evolution of scour around offshore structures. ICE-Maritime Engineering. 2010;163(1):3-17. DOI: 10.1680/maen.2010.163.1.3

65. Hartvig PA, Thomsen JMC, Frigaard P, et al. Experimental study of the development of scour and backfilling. Coastal Engineering Journal. 2010;52(02):157-194. DOI: 10.1142/S0578563410002154

66. Sørensen SPH, Ibsen LB, Frigaard P. Experimental evaluation of backfill in scour holes around offshore monopiles. In: Frontiers in Offshore Geotechnics II. London: Taylor & Francis Group; 2010. pp. 635-640

67. Sørensen SPH, Ibsen LB. Assessment of foundation design for offshore monopiles unprotected against scour. Ocean Engineering. 2013;63:17-25. DOI: 10.1016/j.oceaneng.2013.01.016

68. Reese LC, Cox WR Koop FD. Analysis of laterally loaded piles in sand. In: Proceedings of the Offshore Technology Conference; Houston, TX; 1974. p. 2080

69. O'Neill MW, Murchison JM. An Evaluation of p-y Relationships in Sands, Report to American Petroleum Institute. Houston, TX, USA: University of Houston; 1983

70. Suryasentana SK, Lehane BM. Numerical derivation of CPT-based p-y curves for piles in sand. Géotechnique. 2014;64(3):186-194. DOI: 10.1680/geot.13.P.026

71. Zhu B, Sun YX, Chen RP, Guo WD, Yang YY. Experimental and analytical models of laterally loaded rigid monopiles with

hardening p-y curves. Journal of Waterway, Port, Coastal, and Ocean Engineering. 2015;141(6):04015007. DOI: 10.1061/(ASCE)WW.1943-5460.0000310

72. American Petroleum Institute. Geotechnical and Foundation Design Considerations, ANSI/API Recommended Practice 2 GEO. 1st ed; 2011

73. Qi WG, Gao FP, Randolph MF, Lehane BM. Scour effects on p-y curves for shallowly embedded piles in sand. Géotechnique. 2016;8:648-660. DOI: 10.1680/jgeot.15.P.157

74. Lin C, Bennett C, Han J, Parsons RL. Scour effects on the response of laterally loaded piles considering stress history of sand. Computers and Geotechnics. 2010;37:1008-1014. DOI: 10.1016/j.compgeo.2010.08.009

75. Lin C, Han J, Bennett C, Parsons RL. Analysis of laterally loaded piles in sand considering scour hole dimensions. Journal of Geotechnical and Geoenvironmental Engineering. 2014;140:04014024. DOI: 10.1061/(ASCE)GT.1943-5606.0001111

76. Yang XF, Zhang CR, Huang MS, Yuan JY. Lateral loading of a pile using strain wedge model and its application under scouring. Marine Georesources & Geotechnology. 2017;36(3):340-350. DOI: 10.1080/1064119X.2017.1317889

77. Tseng WC, Kuo YS, Chen JW. An investigation into the effect of scour on the loading and deformation responses of monopile foundations. Energies. 2017;10(8):1190. DOI: 10.3390/en10081190

78. Sørensen SPH. Soil-structure interaction for non-slender, large-diameter offshore monopiles [PhD thesis]. Aalborg, Denmark: Aalborg University; 2012

79. Lin C. Wu R. Evaluation of vertical effective stress and pile lateral capacities considering scour-hole dimensions. Canadian Geotechnical Journal. 2018;56(1):135-143. DOI: 10.1139/cgj-2017-0644

80. Bekken L. Lateral behavior of large diameter offshore monopile foundations for wind turbines [PhD thesis]. TU Delft: Delft University of Technology; 2009

81. Byrne BW, McAdam R, Burd HJ, et al. New design methods for large diameter piles under lateral loading for offshore wind applications. In: Proceedings of 3rd International Symposium on Frontiers in Offshore Geotechnics. Oslo: CRC Press; 2015. pp. 1-6

82. Gao FP, Li JH, Qi WG, Hu C. On the instability of offshore foundations: theory and mechanism. Science China - Physics Mechanics & Astronomy. 2015;58(12):124701. DOI: 10.1007/s11433-015-5745-9

83. Qi WG, Gao FP. Numerical study of local scour effects on the lateral pile-soil interaction. In: Scour and Erosion: Proceedings of the 8th International Conference on Scour and Erosion; Oxford, UK; 12–15 September 2016. pp. 293-300

84. Poulos H, Hull T. The role of analytical geomechanics in foundation engineering. In: Foundation Engineering: Current Principles and Practices. Reston, VA: ASCE 2; 1989. pp. 1578-1606

85. Zhang CR, Zhang X, Huang MS, Tang HW. Responses of caisson-piles foundations to long-term cyclic lateral load and scouring. Soil Dynamics and Earthquake Engineering. 2019;119:62-74. DOI: 10.1016/j.soildyn.2018.12.026

86. Achmus M, Kuo YS, Abdel-Rahman K. Numerical investigation of scour effect on lateral resistance of windfarm monopiles. In: The Twentieth International Offshore and Polar Engineering Conference. Beijing, China: International Society of Offshore and Polar Engineers; 2010. pp. 619-623

87. Bisoi S, Haldar S. Dynamic analysis of offshore wind turbine in clay considering soil-monopile-tower interaction. Soil Dynamics and Earthquake Engineering. 2014;63:19-35. DOI: 10.1016/j.soildyn.2014.03.006

88. Shadlou M, Bhattacharya S. Dynamic stiffness of monopiles supporting offshore wind turbine generators. Soil Dynamics

and Earthquake Engineering. 2016;88:15-32. DOI: 10.1016/j.soildyn.2016.04.002

89. Kallehave D, Byrne BW, LeBlanc Thilsted C, Mikkelsen KK. Optimization of monopiles for offshore wind turbines. Philosophical Transactions of the Royal Society A - Mathematical Physical and Engineering Sciences. 2015;373:20140100. DOI: 10.1098/rsta.2014.0100

90. Prendergast LJ, Gavin K, Doherty P. An investigation into the effect of scour on the natural frequency of an offshore wind turbine. Ocean Engineering. 2015;101:1-11. DOI: 10.1016/j.oceaneng.2015.04.017

91. Prendergast LJ, Reale C, Gavin K. Probabilistic examination of the change in eigen frequencies of an offshore wind turbine under progressive scour incorporating soil spatial variability. Marine Structures. 2018;57:87-104. DOI: 10.1016/j.marstruc.2017.09.009

92. Damgaard M, Andersen LV, Ibsen LB. Dynamic response sensitivity of an offshore wind turbine for varying subsoil conditions. Ocean Engineering. 2015;101:227-234. DOI: 10.1016/j.oceaneng.2015.04.039

93. Damgaard M, Ibsen LB, Andersen LV, Andersen JK. Cross-wind modal properties of offshore wind turbines identified by full scale testing. Journal of Wind Engineering and Industrial Aerodynamics. 2013;116:94-108. DOI: 10.1016/j.jweia.2013.03.003

94. Schafhirt S, Page A, Eiksund GR, Muskulus M. Influence of soil parameters on the fatigue lifetime of offshore wind turbines with monopile support structure. Energy Procedia. 2016;94:347-356. DOI: 10.1016/j.egypro.2016.09.194

95. Rezaei R, Duffour P, Fromme P. Scour influence on the fatigue life of operational monopile-supported offshore wind turbines. Wind Energy. 2018;21(9):683-696. DOI: 10.1002/we.2187

96. Lehane BM, Pedram B, Doherty JA, Powrie W. Improved performance of monopiles when combined with footings for tower foundations in sand. Journal of Geotechnical and Geoenvironmental Engineering. 2014;140:04014027. DOI: 10.1061/(ASCE)GT.1943-5606.0001109

97. Anastasopoulos I, Theofilou M. Hybrid foundation for offshore wind turbines: Environmental and seismic loading. Soil Dynamics and Earthquake Engineering. 2016;80:192-209. DOI: 10.1016/j.soildyn.2015.10.015

98. Stone KJL, Arshi HS, Zdravkovic L. Use of a bearing plate to enhance the lateral capacity of monopiles in sand. Journal of Geotechnical and Geoenvironmental Engineering. 2018;144(8):04018051. DOI: 10.1061/(ASCE)GT.1943-5606.0001913

99. Wang X, Zeng X, Li X, Li J. Investigation on offshore wind turbine with an innovative hybrid monopile foundation: an experimental based study. Renewable Energy. 2019;132:129-141. DOI: 10.1016/j.renene.2018.07.127

100. Whitehouse RJS, Sutherland J, Harris JM. Evaluating scour at marine gravity foundations. In: Proceedings of the ICE-Maritime Engineering; 2011. pp. 143-157

101. Qi WG, Gao FP. Wave induced instantaneously-liquefied soil depth in a non-cohesive seabed. Ocean Engineering. 2018;153:412-423. DOI: 10.1016/j.oceaneng.2018.01.107

102. Qi WG, Li CF, Jeng DS, Gao FP, Liang Z. Combined wave-current induced excess pore-pressure in a sandy seabed: Flume observations and comparisons with theoretical models. Coastal Engineering. 2019;147:89-98. DOI: 10.1016/j.coastaleng.2019.02.006

103. Lin Z, Pokrajac D, Guo Y, Jeng DS, Tang T, Rey N, et al. Investigation of nonlinear wave-induced seabed response around mono-pile foundation. Coastal Engineering. 2017;121:197-211. DOI: 10.1016/j.coastaleng.2017.01.002

104. Zhao HY, Jeng DS, Liao CC, Zhu JF. Three-dimensional modeling of wave-induced residual seabed response around a mono-pile foundation. Coastal Engineering. 2017;128:1-21. DOI: 10.1016/j.coastaleng.2017.07.002

105. Zhu B, Ren J, Ye GL. Wave-induced liquefaction of the seabed around a single pile considering pile-soil interaction. Marine

Georesources & Geotechnology. 2018;36(1):150-162. DOI: 10.1080/1064119X.2017.1340374

106. Duan LL, Jeng DS, Wang D. PORO-FSSI-FOAM: Seabed response around a mono-pile under natural loadings. Ocean Engineering. 2019;184:239-254. DOI: 10.1016/j.oceaneng.2019.05.024

107. Sui TT, Zhang C, Guo YK, Zheng JH, Jeng DS, Zhang JS, et al. Three-dimensional numerical model for wave-induced seabed response around mono-pile. Ships and Offshore Structures. 2016;11(6):667-678. DOI: 10.1080/17445302.2015.1051312

108. Li Y, Ong MC, Tang T. Numerical analysis of wave-induced poro-elastic seabed response around a hexagonal gravity-based offshore foundation. Coastal Engineering. 2018;136:81-95. DOI: 10.1016/j.coastaleng.2018.02.005

05 ANALYTICAL MODEL FOR FROST HEAVE DEFORMATION IN MICROHEAVE FILLER

1. INTRODUCTION: AN OVERVIEW

China now holds the record for having the highest total length of high-speed railway tracks, the quickest speed at which these trains operate, and the largest number of ongoing construction projects in this field. As of the end of 2016, the total length of operational high-speed railway tracks has reached 2.2 × 104 kilometers, making it the longest in the world. High-speed railways possess notable characteristics such as rapidity, comfort, and safety when contrasted with conventional speed railways. In the case of ballasted track, it is important to ensure that the settlement of the high-speed railway subgrade does not exceed 5 or 3 cm in transition portions. In the construction of ballastless track high-speed railway subgrades, the ultimate objective is to achieve zero settlement. This is especially crucial for sections that are prone to deformation, such as transit segments. In these areas, settlement must be limited to less than 5 mm and should not exceed 1/1000 at bends. In order to accomplish the aforementioned objective, it is essential to ensure the smoothness and stability of track structures. This relies on the provision of subgrade structures that possess high strength, high rigidity, consistent longitudinal variation, and superior stability and durability. Presently, in the northeastern region of China, the high-speed trains connecting Harbin-Dalian, Panjin-Yingkou, Harbin-Qiqihar, Shenyang-Dandong, and Jilin-Hunchun traverse areas with seasonal permafrost. In permafrost regions, high-speed trains experience uneven settlement due to subgrade filler frost heave, which is a significant problem when compared to high-speed railways in non-frost regions [1]. All railway subgrades in northern China have experienced different degrees of frost heave, posing a direct risk to travel safety [2]. China adheres to the criteria outlined in the Code for Design of Special Subgrade of Railway (TB10035-2006) when it comes to selecting fillers and designing impermeable and anti-frost-heave constructions for both existing high-speed trains and those currently being built. The accepted norm for controlling filler frost heave is 1%. Nevertheless, in Northeast China's high-speed railway subgrades, the frost can penetrate up to a depth of 3 meters. Based on this criterion, the deformation caused by subgrade frost heave does not satisfy the need for controlling deformation in ballastless tracks. Furthermore, according to the Code for Design of Special Railway Subgrades, the specific grading item for seasonal frozen soil frost heave is naturally occurring frozen soil, rather than artificially compacted subgrade filler. Hence, it is imperative to examine the microheave deformation mechanism of the subgrade filler and develop a precise analysis model to establish a theoretical foundation for the development of specifications and engineering design. At present, researchers are studying subgrade seasonal frost heave issues around the world using experiments and theoretical models. Askar and Zhanbolat conducted a study to examine the impact of freeze-thaw cycles on a high-speed highway subgrade. They performed a frost heave test using soil samples from the field to determine the occurrence of frost heave, the pressure caused by frost heave in the local soil samples, and the depth of frost heave under various conditions. The authors proposed that the problem of frost heave could be mitigated by substituting frost heave susceptible material with coarse-grained soil and implementing a drainage system [3]. Konrad proposed that assessing the anti-frost property of subgrade filler only based on grain grading is insufficient; the frost heave characteristics of fine grains should also be taken into account [4]. Furthermore, the susceptibility of subgrade fillers to frost heave, when they have excellent grading, increases in a linear manner with the amount of fine grains and active minerals present [5]. Bilodeau et al. conducted a study to examine the influence of grain grading and fine grain characteristics on frost heave sensitivity. They discovered that when the fine grain concentration was minimal, the mineral composition and grading of the fine grains had a more significant effect on frost heave sensitivity. Furthermore, the use of fine grains with varying uniformity factors led to varying levels of frost heave [6]. Uthus et al. proposed that the primary determinants of frost heave were water and fine grains, specifically active minerals. Water had a more significant influence on frost heave than the material's elastic modulus [7]. According to Wang et al., the sensitivity of frost heave and the strength of coarse grained soil were influenced by the amount of fine grain content. testing on frost heave and triaxial testing demonstrated that a fine grain soil concentration of 5% resulted in modest frost heave property but strong shear strength for coarse grained soil [8]. Bilodeau et al. examined the correlation between freeze-thaw cycles and the permanent displacement caused by long-term compression in coarse-grained materials in the subgrade. The findings indicated that pressure exerted a more significant impact on a saturated soil sample that had not experienced freeze-thaw cycling. Furthermore, the occurrence of freeze-thaw cycles exerted a notable impact [9]. Guo et al. conducted a frost heave experiment on the Lanzhou-Xinjiang railway subgrade soil. They tested different levels of water content and compactness in both a closed system and an open system. The researchers concluded that the frost heave rate in the open system was significantly higher than that in the closed system, mainly due to water content replenishment [10]. Sheng et al. conducted a model test to investigate the effects of train cycle load on subgrade soil pore hydro-pressure. They found that the water level in the soil rose to the frost heave layer due to the pump suction effect, leading to continuous frost heave in a high-speed railway subgrade [11]. Abdalla et al. expanded a porosity function model, based on plausible assumptions, to simulate a freeze-thaw cycle following a frost heave forecast [12]. Zhang and Michalowski proposed a thermal-hydro-mechanical (THM) model to analyze the behavior of frost heave sensitive soils, specifically viscous soils. This model considers the effects of temperature, hydraulic pressure, and force field changes caused by freeze-thaw cycles, as well as subsequent physical phenomena like expansion and settlement. Additionally, they presented a numerical simulation for the one-dimensional problem [13]. Bronfenbrener analyzed the transmission of heat and matter in a porous media with fine grains, specifically focusing on frost heave during a phase transition. He developed a model for this problem, incorporating a boundary condition that accounts for movement. Validation using test data demonstrated that, up to a specific degree of accuracy, the model was capable of estimating frost heave [14]. Yin developed a predictive model for the rate of frost heave in a seasonal permafrost location with subgrade soil. The model was produced using field data and theoretical analysis, and was then applied to a specific project [15]. This chapter introduces the concept of subgrade microheave filler, which is based on a field survey conducted in a permafrost location and focuses on the features of high-speed railway subgrade filler and frost heave. A formula for calculating frost heave in closed systems using a microheave filler is derived through an indoor test and theoretical research. A model is developed to analyze the deformation caused by frost heave in microheave filler, and its efficacy is demonstrated.

2. EXPERIMENTAL RESEARCH ON THE ELEMENTS THAT INFLUENCE MICROHEAVE FILLER FROST HEAVE

Upon conducting an experimental study, Wang et al. suggested that there were three categories of railway subgrade filler: over coarse grained soils, coarse grained soils and fine grain soils [16]. In this paper, the large amounts of coarse grained soil filler in high-speed railway subgrades in the seasonal permafrost region of China are called microheave filler.

The bed filler used in high-speed railway subgrades in the permafrost regions of China is arranged as follows. A normal embankment bed surface is filled with a layer of 55-cm grading gravel, below which are 5 cm of thick coarse sand and 2.1 m of thick soil in two groups, A and B. In the frozen depth range, the embankment bed is filled with non-frost-heave soil in groups A and B, and soils in groups A, B and C exist below the beds. The top 1.0 m of the embankment bed floor is filled with non-frost-heave filler in groups A and B. Based on the distribution of filler along the line, the lower sections have filler in groups A and B. Low embankment bed surfaces are filled with grading gravel. Non-frost-heave filler in groups A and B is used in the range of the upper meter of bed floors. The lower sections are filled with filler in groups A and B. In the bed floor lower sections, the grading gravel layer water content is 3.2–5.9%, and the fine grain content is 3.2–10.7%. In lower sections, the non-frost-heave soil filler water content generally is 3.7–9.2%, and the fine grain content is 0.4–9.3%. In seasonal permafrost regions, the entire beds are excavated and replaced. The bed surfaces are filled with grading gravel, the upper 1 m of the bed floors are filled with non-frost-heave filler in groups A and B, and the lower sections are filled with filler in groups A and B. The grading gravel layer water contents are 2.2–3.2%, the fine grain contents are 8.0–15.1%, the water contents in the lower section non-frost-heave soil filler are generally 3.7–9.2%, and the fine grain contents are 10.9–16.9%.

Coarse grained soil frost heave tests and analysis proved that the factors that influence major coarse grained soil frost heave include the soil property, water content, temperature and load [17]. To investigate the characteristics of high-speed railway subgrade filler frost heave in permafrost regions, in this chapter, typical permafrost region subgrade soil samples were selected to perform soil property tests for filler covering water content, liquid plastic limit and grain analysis. On this basis, the effects of various factors on filler frost heave were analyzed to obtain rules regarding the influence of soil fine grain content, compactness, water content, permeability and coarse grain pore filling by fine grains on the properties of coarse grained soil frost heave. It is worth noting that the test reported in this section was based on the conventional liquid and plastic limit combined determination method. In the test, fine grain soil liquid limit water and plastic limit water contents were measured to calculate the soil plasticity index and liquid index, which were used as a basis to evaluate the properties of filler soils. A standard screening test was employed to calculate the relative content of each grain group and to determine soil grain composition. In addition, for subgrades, compactness is not only a control index that ensures subgrade filling quality but also a major factor that affects the properties of filler frost heave. Under identical conditions, as compactness increases, fine grain soil frost heave increases initially and then decreases.

3. MICROHEAVE FILLER FROST HEAVE TEST

3.1 Test apparatus

The frost heave test apparatus consisted of a specimen box, an incubator, a temperature control system, a temperature monitoring system, a deformation monitoring system, a pressure loading system and a water supply system, as shown in Figure 1. The specimen box had a diameter of 15 cm and a height of 15 cm.

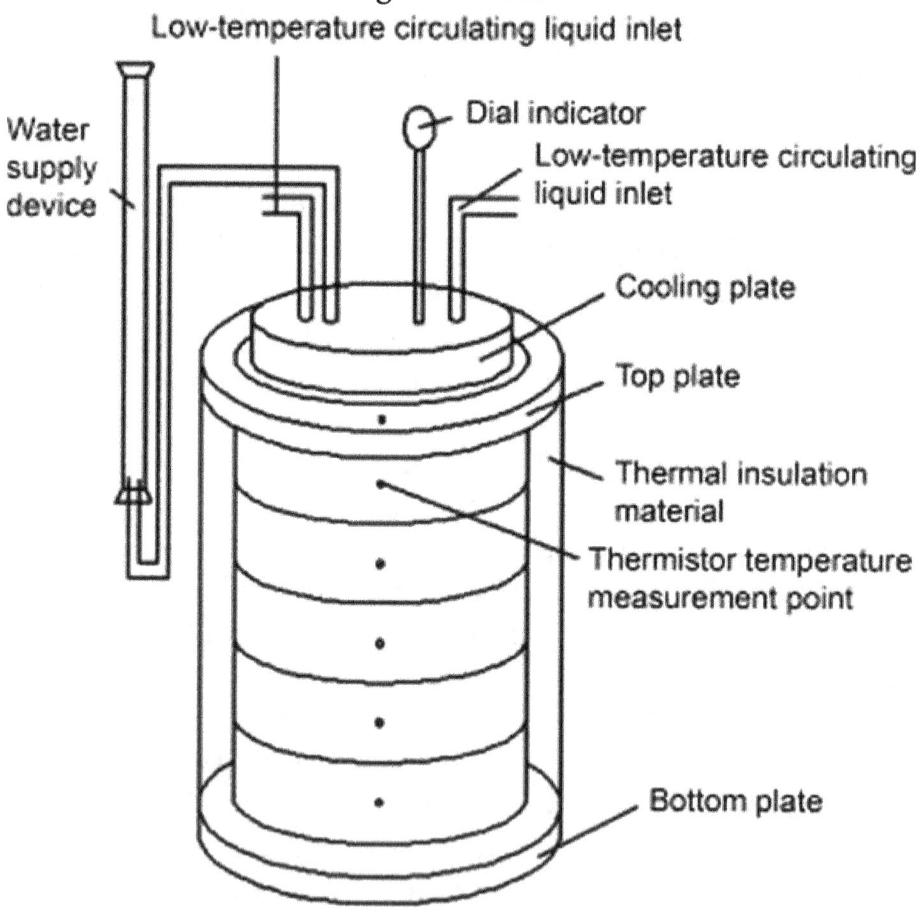

Figure 1. Frost heave test apparatus.

3.2 Test method

Proper amounts of dry soil samples were mixed with water to produce the required water contents. The soil samples were separated into layers of predefined compactness (5 layers, each with thicknesses of 3 cm) and placed in the specimen box. The specimen box with a sample was placed in the incubator. Thermistor thermometers were deployed on the specimen top plate, bottom plate and side. The specimen box was wrapped in 5 cm of thick plastic foam for heat insulation. The top plate and bottom plate temperatures were controlled using two high precision low temperature circulation cooling systems. A dial indicator was installed on the top plate to monitor specimen deformation. Finally, the temperatures for the specimens were collected automatically via the data collector. The testing was performed in a closed environment. The unidirectional freezing method was applied [18], the duration of the freezing process for each specimen was 72 h, and the direction of freezing was from top to bottom. At the beginning of the tests, the soil column temperature was stabilized at approximately 1°C and maintained for 6 h. The bottom plate temperature was then maintained at 1°C. At 0.5 h, the top plate temperature was decreased to -15°C, and the soil samples were frozen rapidly from the top surface. The top plate temperature was increased to -2°C, and top plate temperature was then decreased by a certain amount per hour. At the conclusion of each test, the soil samples were separated into layers in the low temperature incubator, and the water contents were measured.

3.3 Test items

Test filler was obtained from the subgrade filling soil of a high-speed railway in Northeast China. The grades are listed in Table 1. The filler plastic limit ωp was 19.2%, and the optimal water content $\omega 0$ was 13.2%.

Table 1. Test filler grades.

Grain size range (mm)	content (%)
31.5–45.0	20
22.4–31.5	10
7.1–22.4	25

Grain size range (mm)	content (%)
1.7–7.1	20
0.5–1.7	10
0.1–0.5	9
0.075–0.1	3
<0.075	3

To investigate the effects of factors such as water content, filler and external load on microheave filler frost heave for this paper, a frost heave experimental study was conducted to investigate the following aspects.

- **Effects of filler plastic limit and optimal water content on microheave filler frost heave.**

Proper amounts of dry filler were mixed with various amounts of water to produce 30 groups of ω specimens with various levels of initial water content. The initial water content values of the 30 groups of specimens are listed in Table 2. The first 15 groups of specimens were used to investigate the effects of filler plastic limit on the microheave filler frost heave property, and the next 15 groups of specimens were used to investigate the effects of filler optimal water content on the microheave filler frost heave property. The required amounts of the specimens were calculated based on a compactness of 0.95. The specimens were then compacted in the specimen box for frost heave testing.

- **Effects of volumetric water content on microheave filler frost heave**

Table 2. Specimen initial water contents.

Specimen	Initial water content ω(%)
1	12.3
2	14.5
3	15.1
4	15.5
5	15.9

Specimen	Initial water content ω(%)
6	16.0
7	16.5
8	18.0
9	18.5
10	18.8
11	19.0
12	19.6
13	20.0
14	20.5
15	21.3
16	13.3
17	14.3
18	14.4
19	14.7
20	15.5
21	15.6
22	15.8
23	16.2
24	16.3
25	17.5
26	17.8
27	18.0
28	19.1
29	19.4
30	21.7

Proper amounts of dry filler were mixed with various amounts of water to produce 15 groups of ωV specimens with various levels of initial water content. The initial water content values of the 15 groups of specimens are listed in Table 3. The required amounts of the specimens were calculated based on a compactness of 0.95. The specimens were then compacted in the specimen box for frost heave testing.

- **Effects of filler content and filling rate on microheave filler frost heave**

Table 3. Specimen volumetric water contents.

specimen	volumetric water content ωV(%)
1	5.83
2	7.02
3	8.30
4	9.60
5	11.07
6	11.34
7	11.85
8	13.17
9	14.48
10	15.27
11	15.80
12	16.04
13	16.57
14	17.49
15	18.21

To prepare specimens with the required filler contents, proper amounts of dry filler were mixed with the required filler. In addition, the specimen's filler volume fill rate s was calculated via the following formula.

$$s = \frac{1\text{-}V_1}{V_2} \times 100\% \quad (1)$$

where

$$V_1 = \frac{M_1}{\gamma_1},\ V_2 = \frac{M_2}{\gamma_2}.$$

V1 is the filler skeleton grain volume fraction, V2 is the filler volume fraction, M1 is the skeleton grain mass for a unit specimen volume, γ1 is the skeleton grain dry density for the corresponding soil sample compactness, M2 is the filler mass for a unit specimen volume, and γ2 is the filler dry density for the corresponding soil sample compactness.

In this test, 15 groups of specimens were prepared. The specimen filler contents and filler filling rates are listed in Table 4. In the test, fillers with different contents were used to ensure that the filler skeleton component was proportionally adjusted under the premise that the total mass of filler remained unchanged. All the specimens used in the frost heave test had 15% water content and compactness values of 0.95.

Table 4. Specimen filler content and filler filling rate.

Specimen	Filler content (%)	Filler filling rate (%)
1	1.0	0.056
2	2.0	0.118
3	2.5	0.198
4	2.8	0.272
5	3.0	0.316
6	3.2	0.334
7	5.0	0.363
8	5.5	0.381
9	6.0	0.445
10	9.0	0.465
11	11.0	0.529
12	16.0	0.532
13	18.0	0.606
14	40.0	0.859
15	45.0	0.928

- **Effects of filler frost heave on microheave filler frost heave.**

Filler frost heave and pure filler frost heave tests were performed for the 15 groups of filler content specimens prepared in (3) under test conditions of 15% water content and 0.95 compactness [19].

5. Effects of overlying load on microheave filler frost heave.

A specimen with a 10% filler content was prepared for the frost heave test under test conditions in which the initial water content was 15% and the overlying loads were 5, 10, 20, 30, 40, 55, 65 and 80 kPa. Figure 2 shows the loading apparatus used in the frost heave test.

Figure 2. Loading apparatus used for the frost heave test.

4. ANALYSIS OF TEST RESULTS

The microheave filler frost heave rate η was calculated via the following formula:

$$\eta = \frac{\Delta h}{H_f} \times 100\% \tag{2}$$

where Δh is the overall specimen frost heave in mm, and Hf is the frozen depth (excluding frost heave) in mm.

4.1 Effects of water content on microheave filler frost heave

- Relation between filler plastic limit, optimal water content and microheave filler frost heave.

Figure 3 shows the relation between ω-ωp (difference between the filler initial water content ω and filler plastic limit water content ωp) and filler frost heave rate η. Figure 4 shows the relation between ω-ω0 (difference between the filler initial water content ω and filler optimal water content ω0) and filler frost heave rate η.

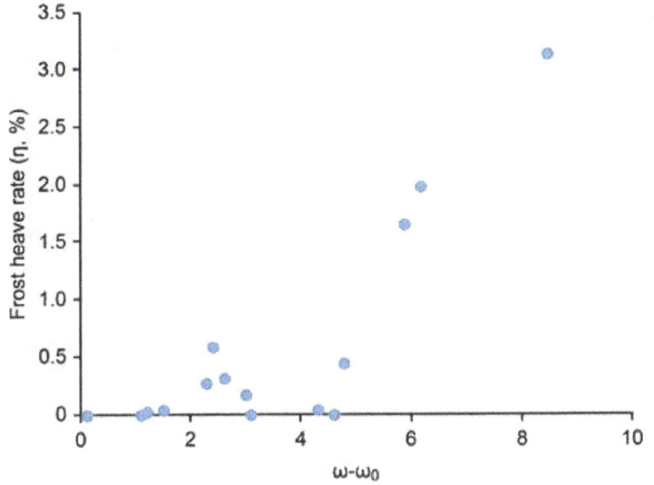

Figure 3. Relation between the filler plastic limit and filler frost heave.

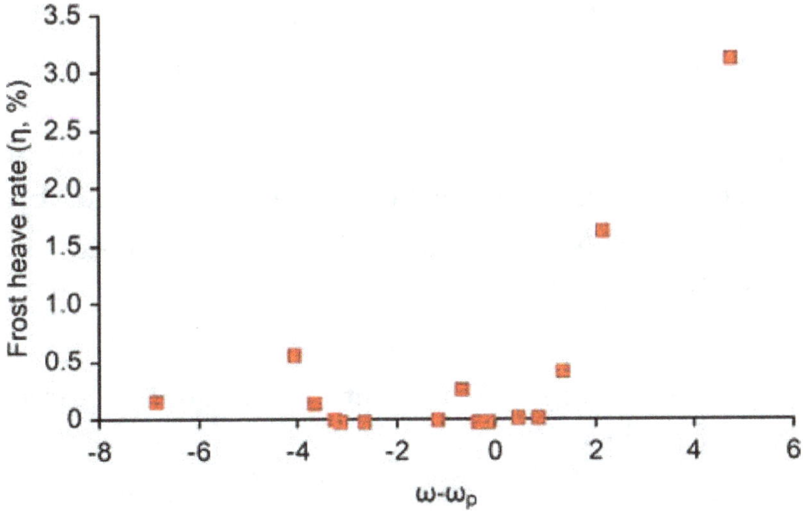

Figure 4. Relation between the filler optimal water content and filler frost heave.

Figures 3 and 4 show that for filler water contents of $\omega \leq \omega p + 2$ or $\omega \leq \omega o + 4.6$, the frost heave rate $\eta < 1\%$. However, as the water content increased further, the filler frost heave rate increased significantly.

- Relation between microheave filler volumetric water content and frost heave rate.

Figure 5 shows the relation between the filler volumetric water content ωv and filler frost heave rate η.

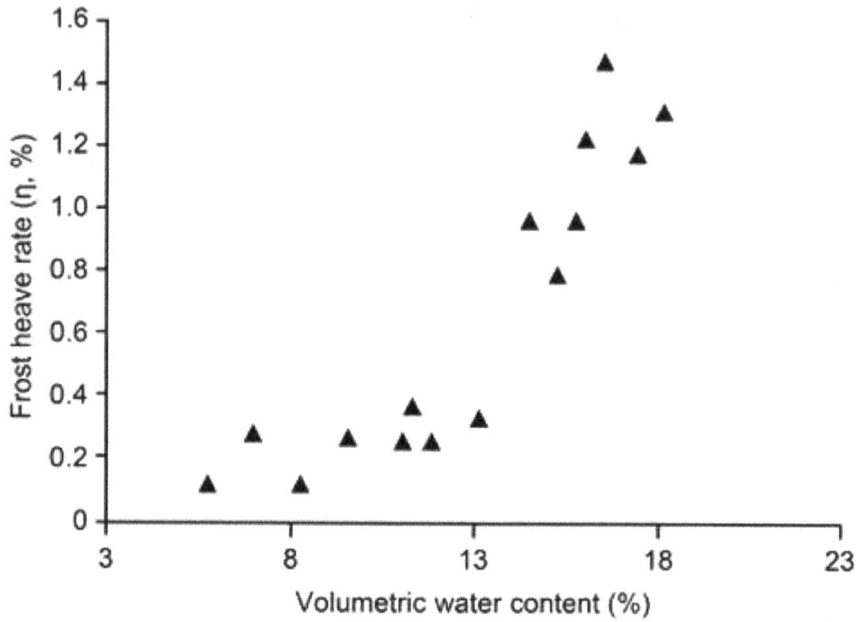

Figure 5. Relation between the filler volumetric water content and filler frost heave rate.

Figure 5 shows that when $\omega v \leq 13\%$, the filler frost heave rate was insensitive to increases in the volumetric water content, and when $\omega v > 13\%$, filler frost heave increased significantly with increasing volumetric water content.

4.2 Effects of filler on microheave filler frost heave

- **Relation between filler content, filling rate and microheave filler frost heave.**

Figure 6 shows the microheave filler frost heave rates for different filler contents.

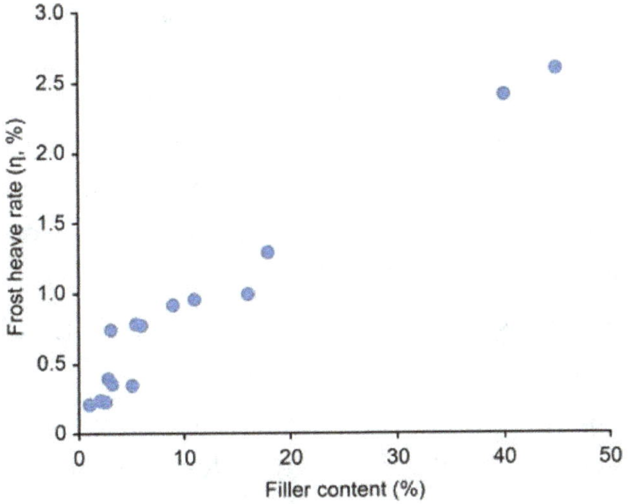

Figure 6. Microheave filler frost heave rate for different filler contents.

Figure 6 shows that filler frost heave increased gradually with filler content. When the filler content was under 3%, the frost heave rate was approximately 0.2%, when the filler content was under 15%, the frost heave rate was under 1.0%, and when the filler content exceeded 15%, the filler frost heave sensitivity increased significantly with filler content.

Figure 7 shows the microheave filler frost heave rates for different filler filling rates.

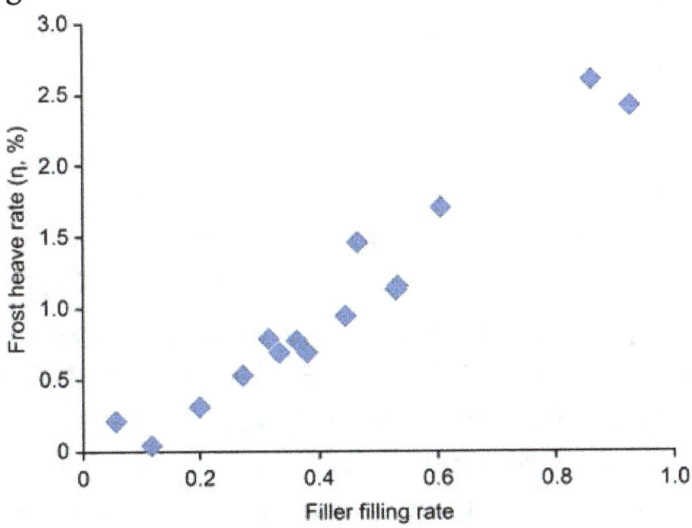

Figure 7. Microheave filler frost heave rate for different filler filling rate.

Figure 7 shows that for filler filling rates below 0.18, the filler frost heave rates were under 0.2%, and when the filler filling rates were under 0.25, the filler frost heave rates were under 0.5%; filler frost heave was insensitive to increases in filler content. For filling rates exceeding 0.25, the filler frost heave rates increased significantly with the filling rate, and frost heave became sensitive. For the filling rates were under 0.37, the frost heave rates were under 1.0%.

- **Relation between filler frost heave and microheave filler frost heave.**

Based on the test results, the relation between specimen filler frost heave and microheave filler frost heave for each group is shown in Figure 8.

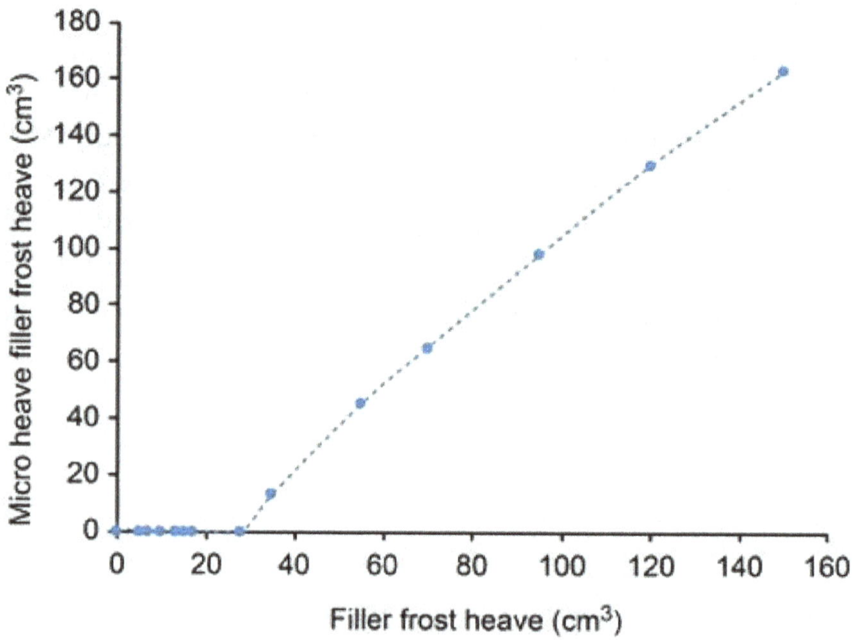

Figure 8. Filler frost heave versus microheave filler frost heave.

Figure 8 shows that when filler frost heave was below 25 cm3, the microheave filler exhibited no frost heave. Under this condition, filler frost heave filled pores, and there was no macroscopic filler frost heave. When filler frost heave exceeded 25 cm3, macroscopic frost heave in the microheave filler began and increased significantly with filler frost heave.

4.3 Relation between overlying load and microheave filler frost heave

Overlying loads affect microheave filler frost heave properties in two ways: the freezing point drops with increasing external load, and overlying loads result in redistributions of filler water content [20]. Figure 9 shows the microheave filler frost heave rates under different overlying loads.

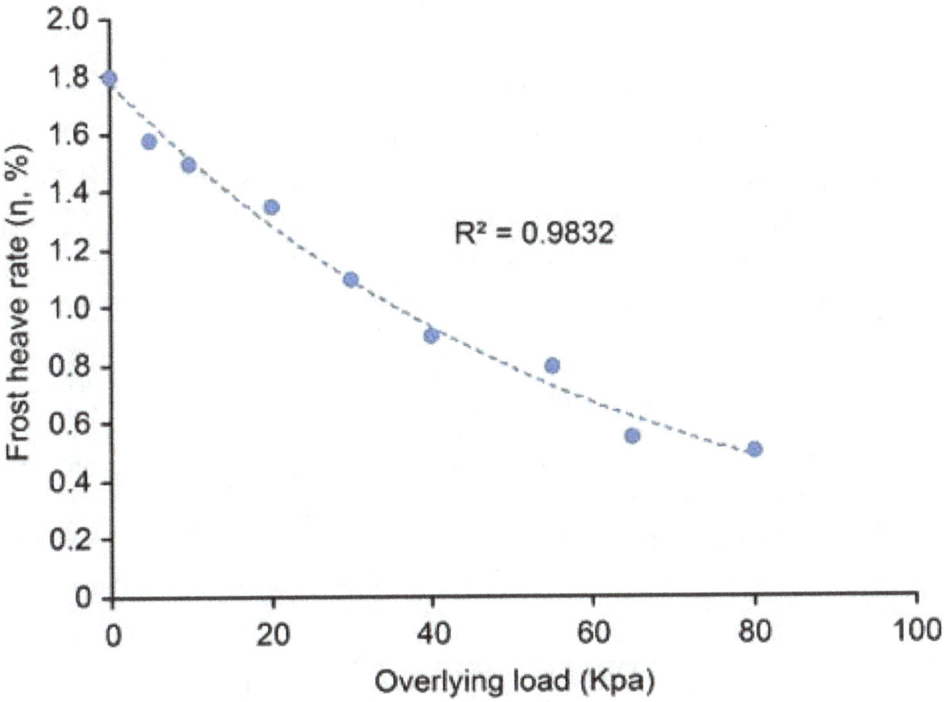

Figure 9. Relation between frost heave rate and overlying load.

Figure 9 shows that overlying load and the microheave filler frost heave rate were exponential related. The microheave filler frost heave rate decreased gradually with increasing overlying load.

The indoor test described above provided data for a qualitative analysis of microheave filler frost heave rules. To further investigate frost heave deformation, theoretical analysis and microheave filler frost heave model creation are required.

5. MICROHEAVE FILLER FROST HEAVE MODEL

5.1 Microheave filler complete filling analysis

When a microheave filler specimen develops frost heave, the coarse grains do not develop frost heave and act as rigid bodies. By comparison, fine grains develop frost heave at low temperatures. Fine grain component volumes consist of soil grain and water content volumes. When fine grains develop frost heave, some water content escapes.

In this section, a simulation embodying the direct expansion of filler grains under natural conditions is examined to study filler grain expansion due to water adsorption. To investigate the complete filling of microheave filler, the mixed material specimen model was simplified. The model was assumed to consist of coarse skeletons, filler grains and pores, as shown in Figures 10 and 11.

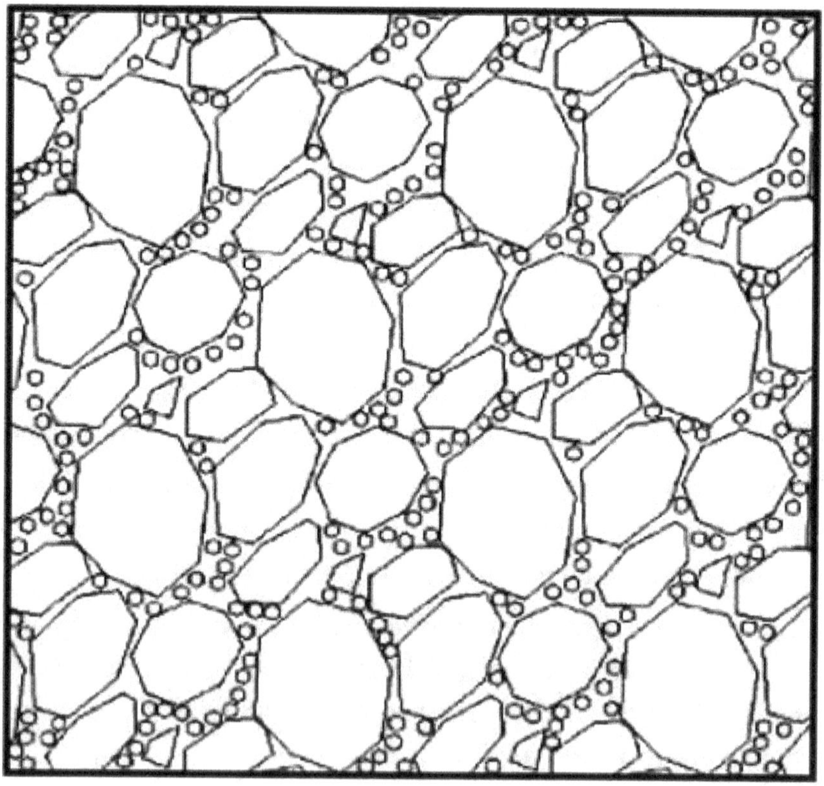

Figure 10. Mixed material.

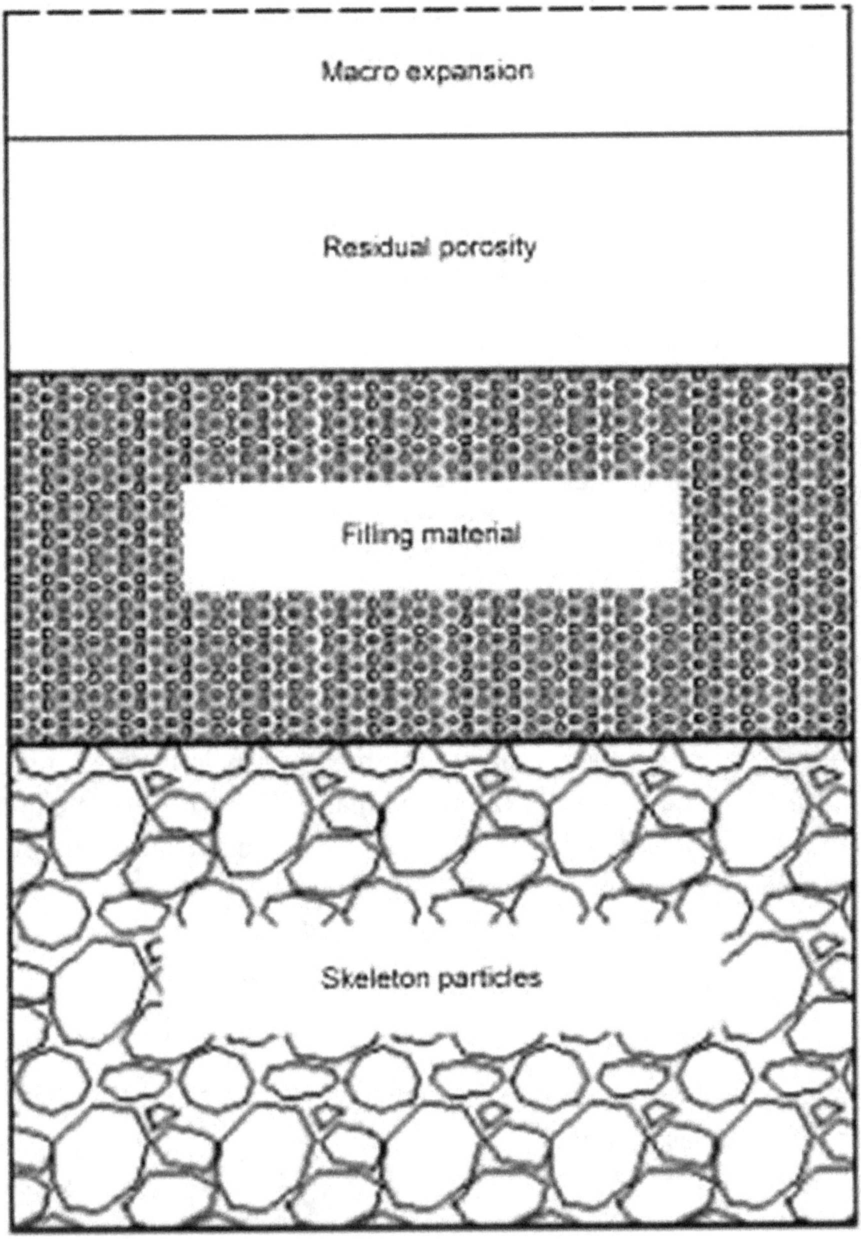

Figure 11. Composition of the mixed material.

Filler and residual pores constituted the mixed material skeleton pores. Among them, filler fills the skeleton pores, expands at low temperature, and, under the side constraint condition, fills the residual pores first. When the residual pores are filled, the expansion is represented as a mixed material specimen macroscopic expansion.

$$\Delta V_{mixed\ material} = \Delta V_{filler} - \Delta V_{skeleton\ pore} \tag{3}$$

If α represents the filler expansion rate, θ represents the filler content, ρmixed represents the mixed material compactness, and Gskeleton represents the skeleton grain ratio, then

$$\Delta V_{skeleton\ pore} = \Delta V_{residual\ pore} \tag{4}$$

From formula (4) and ΔVfiller = Vfillerα,

$$\Delta V_{mixed\ material} = \Delta V_{filler} - \Delta V_{residual\ pore} = \frac{m_{filler}}{\rho_{filler}}(1+\alpha) + \left(\frac{m_{skeleton}}{\rho_{skeleton}}\right) - V_{mixed\ material} \tag{5}$$

Because

$$m_{filler} = \frac{\theta}{(1-\theta)} m_{skeleton} \tag{6}$$

based on the macroscopic volume expansion rate, ηmixed material=ΔVmixed materialπR2Vmixed materialπR2=ΔVmixed material Vmixed material *η*mixed material = Δ*V*mixed material *πR*2*V* mixed material*πR*2 = Δ*V*mixed material *V* mixed material

$$\eta_{mixed\ material} = \frac{\frac{\Delta V_{mixed\ material}}{\pi R^2}}{\frac{V_{mixed\ material}}{\pi R^2}} = \frac{\Delta V_{mixed\ material}}{V_{mixed\ material}} \tag{7}$$

$$\eta_{filler} = \alpha \tag{8}$$

From formulas (7) and (8),

$$\eta_{mixed\ material} = m_{mixed\ material}\left[\frac{\frac{\theta}{(1-\theta)}(1+\alpha)}{\rho_{filler}} + \frac{1}{\rho_{skeleton}}\right] / V_{mixed\ material} - 1 \tag{9}$$

Because

$$\rho_{mixed} = \frac{m_{skeleton} + m_{mixed\ material}}{V_{mixed\ material}} \tag{10}$$

Based on formulas (6) and (10),

$$m_{skeleton} = \rho(1-\theta)V_{mixed\ material} \tag{11}$$

Combining formulas (9) and (11),

$$\begin{aligned} \eta_{mixed\ material} &= m_{mixed\ material}\left[\frac{\frac{\theta}{(1-\theta)}(1+\alpha)}{\rho_{filler}} + \frac{1}{\rho_{skeleton}}\right]/V_{mixed\ material} - 1 \\ &= \rho_{mixed}(1-\theta)\left[\frac{\frac{\theta}{(1-\theta)}(1+\alpha)}{\rho_{filler}} + \frac{1}{\rho_{skeleton}}\right] - 1 \end{aligned} \tag{12}$$

Because the skeleton grain ratio is Gskeleton = ρskeleton/ρw, from formula (12),

$$\eta_{mixed\ material} = \rho_{mixed}(1-\theta)\left[\frac{\frac{\theta}{(1-\theta)}(1+\alpha)}{\rho_{filler}} + \frac{1}{G_{skeleton}\rho_w}\right] - 1 \tag{13}$$

Based on the definition of a mixed material skeleton pore, e = Vresidual pore/VS,

$$\rho_{mixed} = \frac{m_{skeleton} + m_{mixed\ material}}{V_{skeleton} + V_{mixed\ material} + V_{residual\ pore}} = \frac{m_{skeleton} + m_{mixed\ material}}{(1+e)(V_{skeleton} + V_{mixed\ material})} \tag{14}$$

and employing the skeleton grain ratio,

$$\rho_{filler} = \frac{\theta}{\frac{1}{(1+e)\rho_{mixed}} - \frac{(1-\theta)}{(G_{skeleton}\rho_w)}} \tag{15}$$

Based on the definition of the porosity ratio,

$$e = \frac{d_s\rho_w(1+0.001w)}{\rho} - 1 = \frac{d_s\rho_w(1+0.001w)}{\rho_{mixed}} - 1 \tag{16}$$

Based on formulas (15) and (16),

$$\rho_{filler} = \frac{\theta}{\frac{1}{(1+e)\rho_{mixed}} - \frac{(1-\theta)}{(G_{skeleton}\rho_w)}} = \frac{\theta}{\frac{1}{d_s\rho_w(1+0.001w)} - (1-\theta)/(G_{skeleton}\rho_w)} \tag{17}$$

Based on formulas (15) and (17),

$$\eta_{mixed\ material} = \rho_{mixed}(1-\theta)\left[\frac{\frac{\theta}{(1-\theta)}(1+\alpha)}{\rho_{filler}} + \frac{1}{G_{skeleton}\rho_w}\right] - 1$$
$$= \left[\frac{\rho_{mixed}(1+\alpha)}{d_s\rho_w(1+0.001w)} - \frac{\rho_{mixed}(1-\theta)\alpha}{G_{skeleton}\rho_w}\right] - 1 \quad (18)$$

Formula (18) is simplified to

$$\eta_{mixed\ material} = \eta(\rho_{mixed}, \theta, \alpha, d_s, \rho_w, w, G_{skeleton}) \quad (19)$$

Based on formula (19) and an analysis of the situation wherein the microheave filler is completely filled, it was found that mixed material macroscopic expansion is closely related to mixed material compactness ηmixed material, filler content θ, filler expansion rate α, mixed material ratio ds, water compactness ρw, microheave filler water content w and skeleton ratio Gskeleton. Actual frost heave rates are greater than the calculated values.

5.2 Microheave filler partial filling

To further investigate the mixed material frost heave rule, microscopic analysis is required to understand the interactions between pore filler and microheave filler. During filler expansion, mixed material skeleton pores cannot be filled completely. Microheave filler specimens would expand under the side constraint condition. During expansion, the filler would experience plastic deformation and fill skeleton pores, which would result in microheave filler specimen macroscopic elevation. The manifestation of core filler stress is very complex. First, filler grains are not uniformly distributed in all skeleton pores of mixed materials. When stresses on fine aggregates are small, coarse aggregates are not affected, and coarse aggregate skeletons experience no deformation; coarse aggregate and fine aggregate contact surfaces are equivalent to the displacement boundary condition. As the filler expands and fills the remaining part of the skeleton pores, although the fine aggregate stresses are significant, because the expanded filler does not fill all the skeleton pores in the mixed material, the mixed material skeleton grains are not lifted, and the mixed material skeleton does not swell. To summarize, fine aggregates are not completely surrounded by coarse aggregates, pores are interconnected, fine aggregates either expand and deform in large pores or are squeezed to adjacent pores. Therefore, as shown in Figure 12, stress state cannot be described via simple mechanical models.

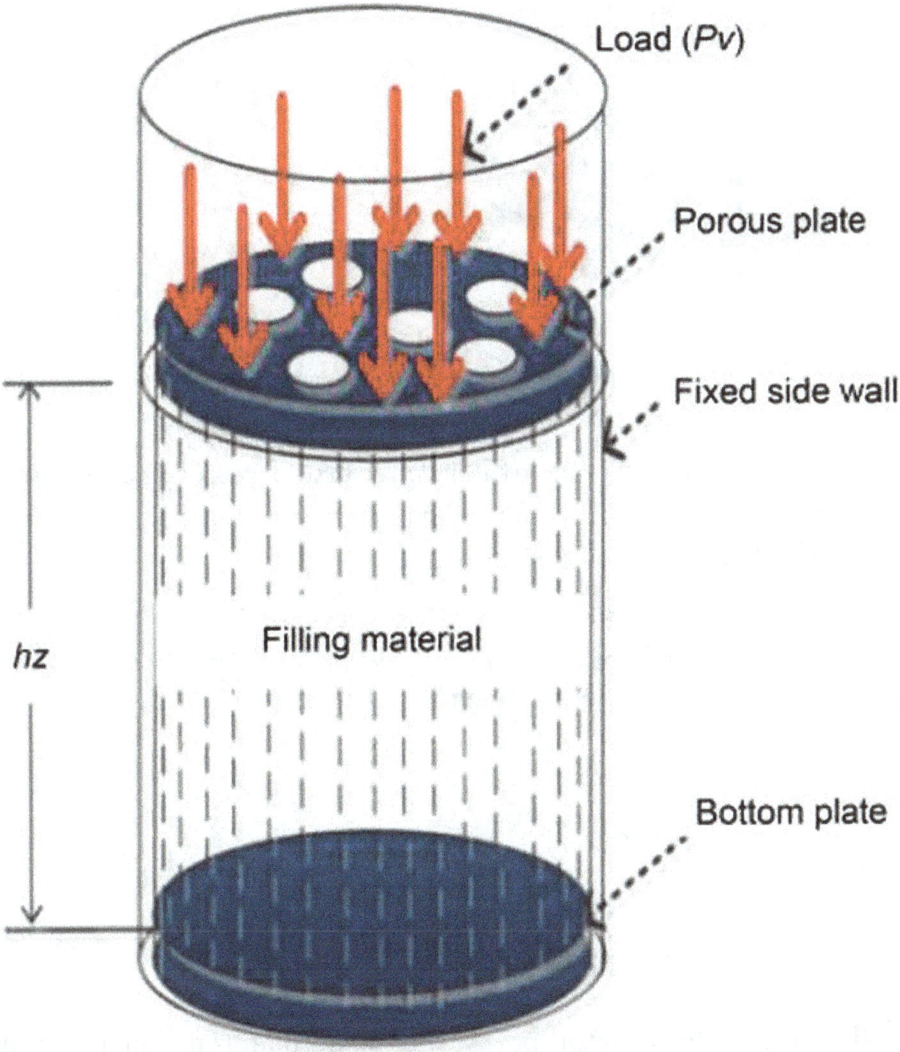

Figure 12. Skeleton pore filled with filler.

Based on Griffith micro-fracture theory and the minimum energy principle, a calculation formula was deduced via mathematical and mechanical analysis to undertake a quantitative analysis of the effects of rock and soil rheological properties for engineering. Filler was pushed into a cylinder using a porous piston. An analogy from physics was applied, and the filler expansion stress status was compared to the mathematical physics model in Figure 13.

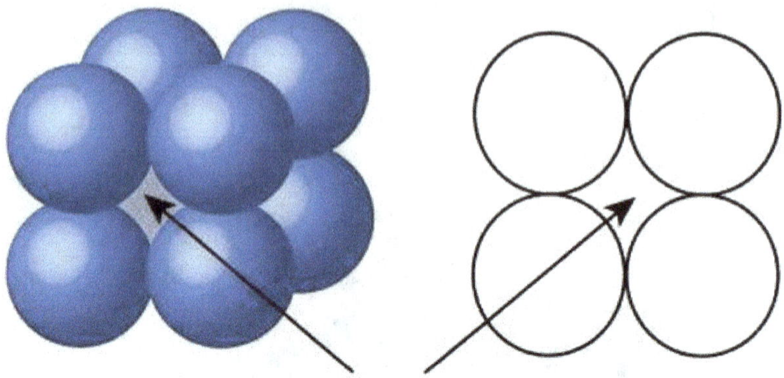

Pores in the framework of particle accumulation;
Filling material distribution in these pores

Figure 13. Mixed material partial filling.

Each pore in the microheave filler skeleton was treated as a container, as shown in Figure 13. The container top plate was porous, and more holes implied greater skeleton porosity.

The filling rate reflects the filling degree, which is defined as the ratio of filler volume to skeleton pore volume. Assuming that the mixed material skeleton grain pore filling rate by filler grains is β,

$$\beta = \frac{V_{filler}}{V_{skeleton\ pore}} \tag{20}$$

For an initial filling rate β0, employing formula (20) yields

$$V_{skeleton\ pore} = \frac{V_{filler}}{\beta} \tag{21}$$

When filler fills skeleton pores, the mixed material skeleton grain volume does not change during expansion. Mixed material macroscopic volume changes can only be achieved by changing the skeleton pores, as shown by formula (22).

$$\Delta V_{mixed\ material} = \Delta V_{skeleton\ pore} = \frac{\Delta V_{filler}}{\beta} + \frac{V_{filler}}{\beta} - \frac{V_{filler}}{\beta_0} \tag{22}$$

When the filler grains expanded, the porous plate moved upward, which increased the macroscopic volume. In addition, some fine grains were extruded via the top plate pore, and plastic deformation consumed energy, as shown by formula (23).

$$\Delta V_{filler} = \Delta V_{mixed\ material} + \Delta V_{extrude} \tag{23}$$

Based on formulas (22) and (23),

$$\Delta V_{extrude} = \Delta V_{filler} - \frac{\Delta V_{filler}}{\beta} - \frac{V_{filler}}{\beta} + \frac{V_{filler}}{\beta_0} \tag{24}$$

Because filler grains are in a plastic state, their elastic potential energies cannot be increased. Therefore, ΔEelastic potential energy = 0.

The mixed material expands by ΔVall, the upper load is N, and the pressure is pv = N/S2. Assume that the piston area is S=S1 + S2, where S1 is the piston area with pores, and S2 is the piston area without pores. Therefore, the work during expansion is.

$$W_{work} = \Delta V_{all}\, p_v \tag{25}$$

Note: ΔVall pv has units of energy, whereas ΔVall N does not have units of energy.

The extrusion volume is calculated via formula (24). The corresponding soil develops plastic deformation and dissipates energy. Based on a general expression of the plastic strain flow rule under the principle of minimum energy consumption, the plastic deformation energy consumption rate is

$$\psi = \int_v \rho\varphi dv \tag{26}$$

Where ρφ = d p. The energy dissipation process only occurs when the yield criterion is satisfied. In this paper, plastic energy is the product of the stress friction coefficient and volume strain. That is, Wplastic deformation energy consumption is proportional to the soil internal friction coefficient K, soil stress P and soil extrusion volume ΔVextrude. In addition, when β is larger and the area of pores in the porous plate is smaller, it is more difficult to extrude soil, and energy consumption is higher. Therefore, it should be positively correlated with 1/(1-β). Plastic energy consumption is described by formula (27):

$$W_{plastic\ deformation\ energy\ consumption} = \frac{\chi KP\Delta V_{extrude}}{(1-\beta)} \tag{27}$$

where K is a dimensionless variable, P is stress, χ is a constant, and χ = (1-β0)b.

The solution for P now follows.

A microelement under a porous plate hole to be extruded is in a plastic flow state, and 3 = 0. 1 = 2 > 0. Based on the Mohr-Coulomb criterion, 1 = 2 = 2ccosφ/(1-sinφ).

Therefore,

$$P = \frac{\sigma_1 + \sigma_2 + \sigma_3}{3} = \frac{4}{3}c\frac{\cos\varphi}{1-\sin\varphi} \tag{28}$$

K represents soil friction, and K = tanφ.

In this process, the expansion potential energy is

$$\begin{aligned} \Delta H &= \Delta E_{elastic\ potential\ enegy} + W_{work} + W_{plastic\ deformation\ energy\ consumption} \\ &= \Delta V_{mixed\ material} P_v + \chi KP\Delta V_{extrude}\frac{1}{1\text{-}\beta} \end{aligned} \tag{29}$$

Based on formulas (22), (24) and (29),

$$\begin{aligned} \Delta H = &\left(\frac{\Delta V_{mixed\ material}}{\beta} + \frac{V_{mixed\ material}}{\beta} - \frac{V_{mixed\ material}}{\beta_0}\right)P_v \\ &+ \chi KP\left(\Delta V_{mixed\ material} - \frac{\Delta V_{mixed\ material}}{\beta} - \frac{V_{mixed\ material}}{\beta} + \frac{V_{mixed\ material}}{\beta_0}\right)\frac{1}{1\text{-}\beta} \end{aligned} \tag{30}$$

According to the principle of minimum energy, the total internal energy of a closed system with steady volume, external parameters and entropy will tend to decrease. When a balance state is reached, the overall internal energy reaches the minimum level. β should result in a minimum ΔH; the range of β is β0 ≤ β ≤ 1.

The prerequisite for minimum ΔH is

$$\frac{\partial \Delta H}{\partial \beta} = 0 \tag{31}$$

The filler grain volume expansion rate is α = ΔVmixed material/ Vmixed material, and b = pv/KP; then, based on (31),

$$\left(\frac{1}{1-\beta}\chi - b\right)\left(\frac{\alpha}{\beta^2} + \frac{1}{\beta^2}\right) + \chi\left(\alpha - \frac{\alpha}{\beta} - \frac{1}{\beta} + \frac{1}{\beta_0}\right)\frac{1}{(1-\beta)^2} = 0 \tag{32}$$

When α = 0, β = β0. This is substituted into the above formula, yielding $\frac{1}{1-\beta_0}\chi - b = 0$, where $\chi = (1-\beta 0)b$.

It is easy to prove that when →+∞, $\beta \rightarrow 1-$. The solution is

$$\left(\alpha\chi - b - ab + \frac{\chi}{\beta_0}\right)\beta^2 + (2ab - 2\chi - 2\alpha\chi + 2b)\beta + (\alpha\chi - b - ab + \chi) = 0 \tag{33}$$

Because the range for β is β0 ≤ β ≤ 1, the solution for the single variable quadratic equation (33) is

$$\beta = \frac{A \pm \sqrt{B - C \cdot D}}{E} \tag{34}$$

where

$$\begin{aligned}
A &= -(2ab - 2\chi - 2\alpha\chi + 2b),\\
B &= (2ab - 2\chi - 2\alpha\chi + 2b)^2,\\
C &= 4\left(\alpha\chi - b - ab + \frac{\chi}{\beta_0}\right),\\
D &= (\alpha\chi - b - ab + \chi),\\
E &= 2\left(\alpha\chi - b - ab + \frac{\chi}{\beta_0}\right).
\end{aligned} \tag{35}$$

Based on formula (34), this is simplified to

$$\beta = \beta(\alpha, b, \chi, \beta_0) \tag{36}$$

where $b=pv/KP$, $pv=N/S=N1+e$, and $\chi=(1-\beta0)b$.

Formula (36) is further optimized:

$$\beta = \beta(N, \alpha, C, \varphi, \beta_0) \tag{37}$$

The formula for calculating the partial filling model expansion filling rate shows that the expansion filling rate is closely related to parameters such as the overlying load, filler cohesion, friction angle and initial fill rate.

The relations between these physical variables are

$$e_{skeleton0} = \frac{V_{skeleton\ pore}}{V_{skeleton\ \mathrm{solid}}} \tag{38}$$

$$\beta_0 V_{skeleton\ pore} = V_{filler} \tag{39}$$

During expansion,

$$\begin{aligned} e_{skeleton} &= \frac{V_{skeleton\ pore} + \Delta V_{skeleton\ pore}}{V_{skeleton\ \mathrm{solid}}} = e_{skeleton0} + \frac{\Delta V_{skeleton\ pore}}{V_{skeleton\ \mathrm{solid}}} \\ &= e_0 + \frac{\Delta V_{filler}}{V_{skeleton\ \mathrm{solid}}} \left(\frac{\alpha}{\beta} + \frac{1}{\beta} - \frac{1}{\beta_0} \right) \end{aligned} \tag{40}$$

As the skeleton grain pore increases, ΔVskeleton pore > 0, eskeleton increases accordingly.

Based on formula (40), if there is only an elevation effect, β0 = β, the ratio of ΔVskeleton pore versus ΔVfiller does not change, and the following always holds:

$$V_{skeleton\ pore} + \Delta V_{skeleton\ pore} = \frac{(V_{filler} + \Delta V_{filler})}{\beta_0} \tag{41}$$

Ease of extrusion is directly related to β. In formula (40), if ΔVfiller is fixed and only β increases, then ΔVskeleton pore decreases; i.e., small extrusion holes and lower elevations make extrusion difficult.

During filler grain expansion, elevation is calculated via formula (31), and extrusion is calculated via formula (32). The ratio of the two is calculated by substituting β into the two formulae.

5.3 Creation of a microheave filler frost heave model

In an actual project, when the filler is expanding, microheave filler skeleton pores cannot be filled completely. Therefore, the microheave filler frost heave complete filling model should be modified. The microheave filler complete filling theoretical model and partial filling model are combined, and interactions between the filling and elevation effects are considered. The model is created using the principle of minimum energy. For complete filling, the microheave filler frost heave model is described by formula (3). For partial filling, the microheave filler model is described by formula (23).

$$\Delta V_{mixed\ material} = \Delta V_{filler} - \left(\frac{\alpha - \frac{\alpha}{\beta} - \frac{1}{\beta} + \frac{1}{\beta_0}}{\frac{1}{\beta} - 1}\right)\left(V_{skeleton\ pore} - V_{filler}\right) \tag{42}$$

Where

$$M = \frac{\alpha - \frac{\alpha}{\beta} - \frac{1}{\beta} + \frac{1}{\beta_0}}{\frac{1}{\beta} - 1} \tag{43}$$

The microheave filler frost heave model is simplified to

$$\Delta V_{mixed\ material} = \Delta V_{filler} - M\left(V_{skeleton\ pore} - V_{filler}\right) \tag{44}$$

The above formula is quantified. Based on formulas (4) and (44),

$$\begin{aligned} \Delta V_{mixed\ material} &= \Delta V_{filler} - M V_{residual\ pore} \\ &= \frac{m_{filler}}{\rho_{filler}}(M + \alpha) + M\frac{m_{keleton}}{\rho_{keleton}} - M V_{mixed\ material} \end{aligned} \tag{45}$$

Based on formulas (3), (45) is rearranged to

$$\Delta V_{\text{混合料}} = m_{\text{骨架}}\left[\frac{\frac{\theta}{1-\theta}(M+\alpha)}{\rho_{\text{充填料}}} + \frac{M}{\rho_{\text{骨架}}}\right] - MV_{\text{混合料}} \tag{46}$$

Based on formulas (7), (8) and (46),

$$\eta_{mixed\ material} = \frac{\Delta V_{mixed\ material}}{V_{mixed\ material}} = m_{skeleton}\left[\frac{\frac{\theta}{(1-\theta)}(M+\alpha)}{\rho_{filler}} + \frac{M}{\rho_{skeleton}}\right]/V_{mixed\ material} - M \quad (47)$$

Substitution of formulas (11) into (47) yields

$$\begin{aligned}\eta_{mixed\ material} &= m_{skeleton}\left[\frac{\frac{\theta}{(1-\theta)}(M+\alpha)}{\rho_{filler}} + \frac{M}{\rho_{skeleton}}\right]/V_{mixed\ material} - M \\ &= \rho_{mixed}(1-\theta)\left[\frac{\frac{\theta}{1-\theta}(M+\alpha)}{\rho_{filler}} + \frac{M}{\rho_{skeleton}}\right] - M\end{aligned} \quad (48)$$

Based on the skeleton grain ratio definition, (49) is rearranged as

$$\begin{aligned}\eta_{mixed\ material} &= \rho_{mixed}(1-\theta)\left[\frac{\frac{\theta}{1-\theta}(M+\alpha)}{\rho_{filler}} + \frac{M}{\rho_{skeleton}}\right] - M \\ &= \rho_{mixed}(1-\theta)\left[\frac{\frac{\theta}{1-\theta}(M+\alpha)}{\rho_{filler}} + \frac{M}{G_{skeleton}\rho_w}\right] - M\end{aligned} \quad (49)$$

Based on formula (17), the left side is ΔVall = ηmixed hS. Based on the complete filling model, hS is 1. Rearranging formula (50) yields

$$\begin{aligned}\eta_{mixed\ material} &= \rho_{mixed}(1-\theta)\left[\frac{\frac{\theta}{1-\theta}(M+\alpha)}{\rho_{filler}} + \frac{M}{\rho_{skeleton}}\right] - M \\ &= \left[\frac{\rho_{mixed}(M+\alpha)}{d_s\rho_{filler}(1+0.01w)} + \frac{\rho_{mixed}(1-\theta)\alpha}{G_{skeleton}\rho_w}\right] - M\end{aligned} \quad (50)$$

where M is given by formula (43), and β in M is given by formula (35).

To determine β, b = pv/KP is used, where K represents soil friction, K = tanφ, and P is determined using formula (33).

The pressure is pv = N/S=N(1 + e), and χ is a constant, where χ = (1-β0) b. Formula (51) is simplified to

$$\eta_{mixed\ material} = \eta(N, \varphi, C, \alpha, \beta_0, d_s, \rho_w, w, \rho_{mixed}, \theta, G_{skeleton}) \quad (51)$$

The parameters in relation formula (51) were obtained via the indoor filler frost heave test, and the simple physical and mechanical test was used for the microheave filler.

Based on the microheave filler frost heave model, microheave filler frost heave is closely related to the specimen overlying load, filler plastic deformation friction angle, filler cohesion, filler frost heave rate, mixed material skeleton pore filling rate by the initial filler, mixed material ratio, water density, mixed material water content, mixed material density, filler content and mixed material skeleton ratio.

6. MODEL VERIFICATION

Based on the model presented in this paper, the filler frost heave rates for various filler filling rates are shown in Figure 14, and the filler frost heave rates for various water contents are shown in Figure 15. These results were then compared with the test results presented in this paper.

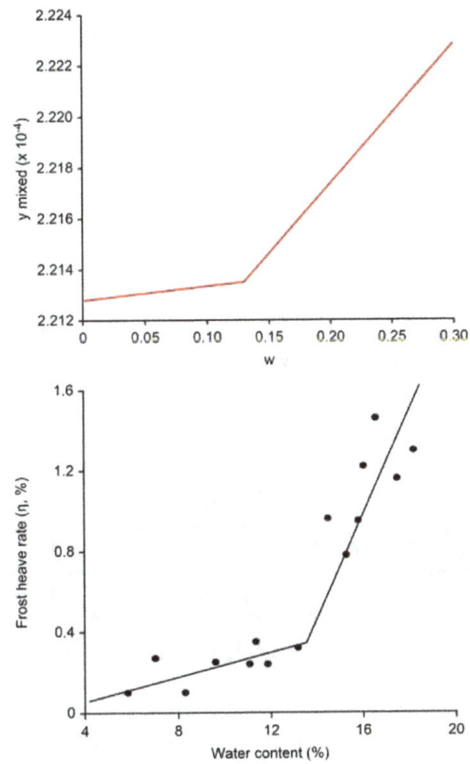

Figure 14. Frost heave rate versus filler content.

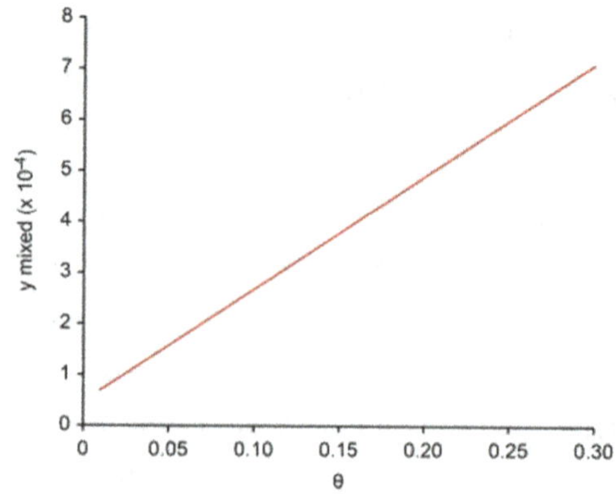

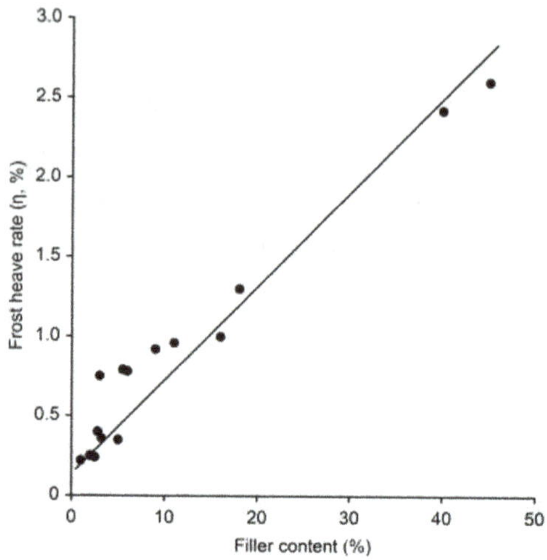

Figure 15. Variations in frost heave rate with water content.

Figures 14 and 15 show that the model simulation results essentially matched the test results. However, the specimens in the test and simulation differed. A silt filler frost heave rate test under an overlying load of N1 = 60 kPa was performed. Other parameters, including the model parameters, are listed in Table 5.

Table 5. Model parameters.

wp	*C1*	φ	ρ	*e*	β0	w	α
9	6	30°	2.1	0.3	0.178	9%	0.22%

The filler frost heave rate in the lab is ηmixed material = 0.123 mm/150 mm = 0.082%.

The simulation showed that when the water content was 9%, the frost heave rate was approximately 0.22%.

A comparison of the calculated and experimental filler frost heave rates showed a difference of only 0.06%, which indicates that the microheave filler frost heave computational model derived in this paper is viable for frost heave rate calculations.

7. CONCISE SUMMARY

To address the issue of seasonal permafrost region subgrade filler microheave in China, an indoor frost heave characteristics test was used to investigate the effects of influencing factors such as filler content, natural water holding capacity, filler grading, permeability coefficient, and water supply condition on the subgrade frost heave rate with and without external constraints (overlying train loads). On that basis, models for complete and partial filling microheave filler frost heave were developed. Indoor testing confirmed the validity of the theoretical microheave filler frost heave rate forecast model. The detailed findings are as follows.

- The filler grading frost heave test showed that when the gravel or finer aggregate was reduced, the filler frost heave rate decreased gradually; however, even for very low levels of fine grain content at low temperatures, the soil still produced a certain level of frost heave. The filler overlying load test showed that an overlying pressure produced a certain level of suppression effects on filler frost heave.
- Microheave filler frost heave increased with filler frost heave; however, when the filler fill rate was under 0.25, as the filler content increased, the changes in filler frost heave were insignificant. This means that at this condition, filler frost heave filled pores, no filler macroscopic frost heave occurred, and soil was insensitive to frost heave. For filler filling rates exceeds 0.25, as the filling rate increased, the filler frost heave rate increased significantly, and the soil was sensitive to frost heave.
- A microheave filler frost heave model, ηmixed material = η(N, φ, C, α, β0, ds, ρw, w,ρmixed, θ, Gskeleton), The overlying load, filler plastic deformation friction angle, filler cohesion, filler

frost heave rate, mixed material skeleton pore filling rate by initial filler, mixed material ratio, water density, mixed material water content, mixed material density, filler content, and mixed material skeleton ratio were all taken into account. Testing confirmed the applicability of the microheave filler frost heave model.

AT A GLANCE

The rapid advancement of high-speed railway systems has introduced new demands for controlling subgrade frost heave deformation. The use of standard non-frost-heave filler does not satisfy the demands of high-speed railways for optimal smoothness and stability, hence posing a risk to the safety of high-speed train operations. This research conducted laboratory tests and theoretical analysis to address the issue of microheave in the subgrade filler of seasonal permafrost regions in China. The study focused on analyzing the physical qualities of components that influence frost heave in the filler. The impact of skeleton grain on the production of frost heave is uncovered. An investigation is conducted on the mechanism of frost heave formation in microheave filler. Using the idea of minimal energy, a calculation formula for frost heave in microheave filler is derived, and a deformation analysis model for frost heave in microheave filler is developed. Furthermore, the model's efficacy is showcased in an indoor evaluation. This paper offers a theoretical framework for managing the frost heaving deformation of railway subgrade.

REFERENCES

1. Ye YS, Cai DG, Yan HY, Yao JP, Zhang QL, Cheng AJ. The mechanism analysis of micro frost heave coarse-grained filling material in the high-speed railway. Japanese Geotechnical Society Special Publication. 2016;2(47):1660-1663
2. Chen X, Liu J, Liu H. Frost action of soil and foundation engineering. Beijing: Science Press; 2011 (in Chinese)

3. Askar Z, Zhanbolat S. Experimental investigations of freezing soils at ground conditions of Astana, Kazakhstan. Sciences in Cold and Arid Regions. 2015;7(4):0399-0406

4. Konrad J-M. Freezing-induced water migration in compacted base-course materials. Canadian Geotechnical Journal. 2008;45(7):895-909

5. Konrad JM, Lemieux N. Influence of fines on frost heave characteristics of a well-graded base-course material. Canadian Geotechnical Journal. 2005;42(2):515-527

6. Bilodeau J-P, Dore G, Pierre P. Gradation influence on frost susceptibility of base granular materials. International Journal of Pavement Engineering. 2008;9(6):397-411

7. Uthus L, Hermansson Å, Horvli I, Hoff I. A study on the influence of water and fines on the deformation properties and frost heave of unbound aggregates. In: International Conference on Cold Regions Engineering, pp. 1-13

8. Wang T, Yue Z, Sun T. Influence of fines content on the anti-frost properties of coarse-grained soil. Sciences in Cold and Arid Regions. 2015;7(4):0407-0413

9. Bilodeau JP, Doré G, Poupart J. Permanent deformation of various unbound aggregates submitted to seasonal frost conditions. Cold Regions Engineering. 2012:155-164

10. Guo Y, Yang Y, Zeng L. Study on frost heaving characteristics of subgrade soil of lanzhou-xinjiang railway. Subgrade Engineering. 2010;2:149-151

11. Sheng DC, Zhang S, Niu F, Cheng G. A potential new frost heave mechanism in high-speed railway embankments. Géotechnique. 2014;64(2):144-154

12. Abdalla B, Fan C, Mckinnon C, Gaffard V, Audibert-Hayet A, Coche E, et al. Extended porosity rate function for frost heave. In: ASME 2014 33rd International Conference on Ocean, Offshore and Arctic Engineering, ASME. San Francisco, CA; 2010. V010T007A046

13. Zhang Y, Michalowski RL. Multiphysical modeling and numerical simulation of frost heave and thaw settlement. In: Geo-congress 2014 Technical papers@ sgeo-characterization and modeling for sustainability, ASCE, Atlanta, GA; 2014. pp. 2735-2744

14. Bronfenbrener L. The modelling of the freezing process in fine-grained porous media: Application to the frost heave estimation. Cold Regions Science and Technology. 2009;56(2):120-134

15. Yin C. Study on Seasonal Permafrost Region Subgrade soil Frost Heave Characteristics and Evaluation Index System. Northeast Forestry University; 2014

16. Wang Z, Zhang Q, Ye Y. In-depth study on the classification of filling for railway subgrade. China Railway Science. 2012;33(2):13-20

17. Li A, Niu YH, Niu F, Liu H. Research status of frost heaving properties and controlling measures of coarse grained soil. Journal of Glaciology and Geocryology. 2015;37(1):202-210

18. Du X. Research on Filler Frost Heave and Coarse Grain Skeleton Interaction-Based Micro Heave Filler Frost Heave Mechanism. Beijing: China Academy of Railway Sciences; 2015

19. Tian Y, Liu J, Peng L. Experimental study on frost action of fine-grained soils under dynamic and static loads. Chinese Journal of Geotechnical Engineering. 2010;32(12):1882-1888

20. Zhao H, Yan H, Zhang Q, et al. Study on frost-heave performances of filler consisting of coarse grained soils for subgrade bed in seasonal frozen region. Railway Engineering. 2014;7:92-94

06 ESTIMATING EXCESS PORE PRESSURE GENERATION AND NONLINEAR SITE RESPONSE OF LIQUEFIED AREAS

1. INTRODUCTION: AN OVERVIEW

Liquefaction is a significant and intricate subject that geotechnical engineers encounter in their current work. The Good Friday earthquake in Alaska and the Niigata earthquake in Japan, both occurring in 1964, were the first instances where the catastrophic impact of liquefaction was detected. This included slope failures, bearing failures of foundations and abutments, and other related effects. Since then, extensive research has been conducted on liquefaction, with numerous researchers contributing to a greater understanding of the topic. However, it remains complex in many aspects. The commencement of the study was in the 1960s by Seed and his colleagues. In 1967, Seed and Ldriss conducted a study on clean sand samples to assess the cyclic resistance of soils, which laid the foundation for the concept of liquefaction. In 1968, Seed conducted further research on this topic. In 1971, Seed and Peacock also contributed to the understanding of liquefaction. Other researchers, such as Finn et al., Castro, and Youd, also focused on studying liquefaction during this period. Under undrained conditions, cyclic loading of saturated coarse soils leads to the accumulation of excessive pore pressure. Liquefaction occurs when the extra pore pressure reaches the same level as the total stress, resulting in zero effective stress [10]. This indicates that the earth has lost its ability to support any weight, leading to failures in the soil and/or significant deformations in the foundations of structures. Hence, it is imperative to consider this phenomenon while building a structure in places prone to liquefaction. The pore pressure buildup primarily relies on the transmission of seismic waves through the soil column induced by an earthquake. Loose soils have a tendency to get denser when they are subjected to loading under drained conditions. However, if the conditions are undrained, the soil's compression during cyclic loading is nearly impossible. In this scenario, the limited time available prevents water from exiting the empty spaces, resulting in the seismic stress causing an elevation in pore pressure. The pore pressure gradually increases until the loading process is completed. If the energy generated by the excitation is sufficiently large, the pressure will eventually approach the entire stress level. Due to the absence of sufficient tension, the soil particles will disperse in the water and thus experience a decrease in both strength and stability. Therefore, it is necessary to scrutinize the vulnerability to liquefaction in regions prone to earthquakes, taking into consideration some fundamental factors: Is the soil sufficiently granular and porous to be prone to liquefaction? Does the region have enough activity to generate sufficient energy to liquefy the soil? If liquefaction occurs, does it predominantly happen at shallow depths, thereby significantly impacting the superstructure above it? Alternatively, does it occur at much greater depths, resulting in no discernible ground effects? The following are the inquiries that must be addressed in seismically active areas to implement preliminary measures for potential dangers. While the liquefaction susceptibility is typically linked to in situ test results such as SPT, CPT, Vs, or soil index parameters like relative density, fine content, Atterberg limits, etc., modern technology allows engineers to conduct laboratory cyclic tests to gain a better understanding of how soil behaves under seismic loads. This technology provides engineers with a high level of confidence. Empirical models are commonly used to assess liquefaction due to their simplicity, although they are less precise compared to the actual-scale pore pressure models found by dynamic triaxial tests performed in the laboratory. The complexity of the models used to generate extra pore pressure remains, however, it is the most suitable method for estimating the dynamic behavior of coarse soils. Correlations do not provide any information regarding pore pressure development, therefore the potential ground deformations caused by liquefaction are only an approximation. However, the actual degrees of deformation can be accurately seen in laboratory settings [11]. It is important to mention that the experiments can be conducted using different levels of strain amplitudes, ranging from 10-6 to 100. The laboratory usually employs dynamic triaxial test equipment to determine the cyclic behavior of soils. The study of pore pressure generation models primarily focuses on three distinct approaches: (i) stress-based models, (ii) strain-based models, and (iii) energy-based models. Stress-based models aim to establish a relationship between the excess pore pressure ratio (ru), which is the ratio of excess pore pressure to effective stress, and the number of cycles required to cause soil liquefaction. Lee and Albaisa (12) and Seed et al. (13) examined the correlation between the ru and (N/Nliq), where N represents the number of loading cycles and Nliq represents the number of cycles required to commence liquefaction. Subsequently, Booker et al. [14] conducted research using the same model but proposed alternative constants. In 2008, Polito et al. [15] developed a model that incorporated fine content, relative density, and cyclic stress ratio. Ultimately, Baziar et al. [16] enhanced the formulation by incorporating additional parameters. The strain-based models assess the change in volumetric strain by considering the cyclic shear strain and incorporating specific constants for the clean sand samples [4, 5, 16, 17]. Dobry et al. (18, 19, 20), Vucetic and Dobry (21), and Carlton (22) provided more intricate models that incorporated new parameters, including threshold cyclic shear strain, loading dimension, and fitting parameters related to fine content, shear wave velocity, and so on. Energy-based models primarily analyze stress-strain loops to determine the necessary energy required for estimating liquefaction susceptibility. One of the most commonly utilized models in this context is the Green et al. [23] approach. This work involved conducting laboratory testing at various cyclic stress ratios (CSR) and comparing the results with stress-based models. The specifics of these models and their formulation will be addressed at a later time. As previously stated, the reduction in soil stiffness poses a significant challenge for civil engineers. According to

Kramer et al. [24], it is essential to accurately estimate the behavior of soils under dynamic loads in order to create reliable and cost-effective structures. In addition to laboratory investigations, building rules govern the design of structures to be constructed in certain regions, with the aim of estimating the dynamic loads that will be transmitted through the soil layers to the structure in various seismic situations. The ground response is highly dependent on the dynamic characteristics of the soils and the specific conditions of the site during seismic events. Building codes typically vary the construction of the response spectrum based on the soil class, which is defined by the shear wave velocity of the top 30 meters. This classification fails to account for the influence of local site effects on the ground reaction that the structure would experience. Hence, the reaction particular to the site plays a crucial role in estimating the dynamic loads transmitted to the surface of the soil. Building codes typically establish regulations that require consideration of site reaction, such as the ZF soil class. However, the accurate prediction of the dynamic loads experienced by a structure can be significantly affected by the nonlinearity of soil behavior. The dynamic behavior of soils can be determined in the laboratory by conducting dynamic triaxial tests for medium to high shear strain levels, and resonant column tests for low to medium shear strain levels. The behavior is characterized by the modulus reduction and damping qualities of soil under different shear strains. The values are derived from the stress-strain loops for each cycle in order to construct two significant curves that illustrate the nonlinearity of the soil. This chapter will categorize the liquefaction investigations into two distinct groups: experimental and analytical approaches. The initiation of excess pore pressure building, which leads to liquefaction of soil, may primarily be evaluated in the laboratory using dynamic triaxial test equipment. A total of eighteen distinct dynamic tests were performed at various CSRs (Cyclic Stress Ratios) and loading conditions to acquire the pore pressure generation and verify the accuracy of the existing models in the literature using the collected data. The second phase of the study will specifically examine the non-linear site response of the areas affected by liquefaction, using the existing in situ data. The analysis utilized five distinct boring logs to identify the nonlinear behavior of the soil layers. It is important to mention that the soil columns experienced liquefaction at varying depths for different earthquake scenarios. The ground responses were computed, and then the effectiveness of the construction codes in predicting the spectral response of the soil columns was evaluated.

2. QUANTIFICATION OF THE ADDITIONAL PORE PRESSURE BUILDUP

In this part, clean sand samples were loaded under varying cyclic stresses with different loading patterns using the dynamic triaxial test system equipped at the Eskisehir Osmangazi University Soil Dynamics Laboratory, and it is presented in Figure 1.

Figure 1. Dynamic triaxial test system at the Eskisehir Osmangazi University.

The system is a fully automated pneumatic system that is capable of conducting axial strains up to 10%, and the tests can be handled as stress controlled or strain controlled. A total of 18 tests was completed to justify the performance of the excess pore pressure generation models suggested in the literature.

2.1 Material and method

In order to study liquefaction in the laboratory, standard sand samples are generally chosen for the tests since the cost of the undisturbed sand samples that is only possible by the freezing technique are really high. Therefore, clean Podima sand samples were used in this study. The type of sand is classified as poorly graded sand (SP) according to the unified soil classification system (USCS), and the grain size distribution is shown in Figure 2. The minimum and maximum unit weights of the sand were determined as 15.30 and 17.66 kN/m3 as the standards suggest, respectively. The specific gravity was determined as 2.63.

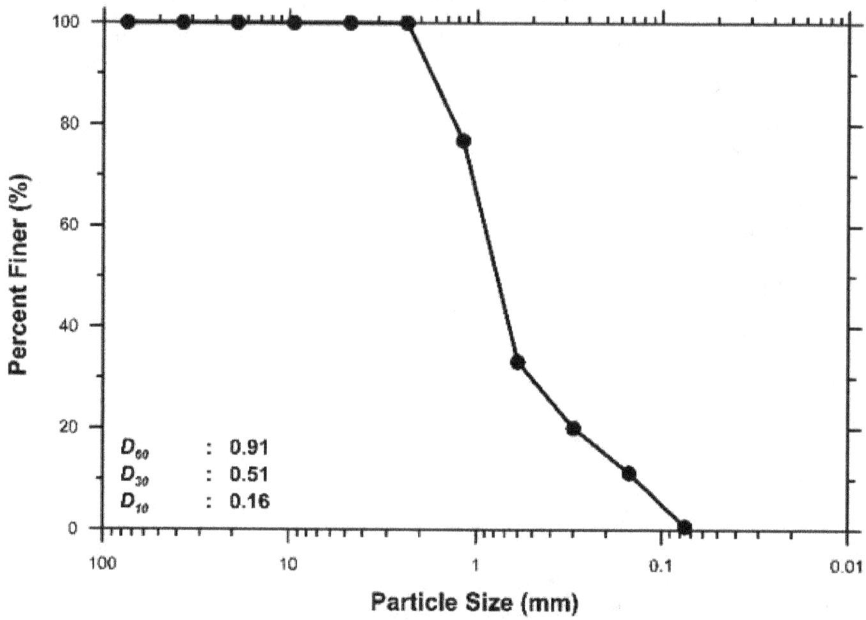

Figure 2. Grain size distribution of the sand used in the study.

The soil samples had a diameter of 7 cm and a height of 14 cm, and the desired relative density was decided to be as 40%. Dried sand was dropped into a two parted mold, which has a membrane placed in, by a funnel with some little shaking and compacting of the sample accordingly. Later, the top header was set on the sample, and some vacuum was applied to the sample in order to hold itself, and the mold was taken apart. Following external cell placement, the cell was filled with water, and some cell and back pressures were applied in such a way that there was a difference of 5 kPa at increasing level. The saturation was controlled with the B constant to reach a value of 0.96. After saturation was accomplished, the sample was consolidated under isotropic condition approximately an hour with a confining pressure of 65 kPa. Finally, soil samples were axially loaded at different amplitude of stresses with changing loading shapes until all the samples were liquefied.

2.2 Test program and loading patterns

In order to understand the excess pore pressure generation at different cyclic stresses, three different stress levels were considered. Cyclic stress ratio is defined as the cyclic stress applied to the sample (either compression or extension) normalized by the confining pressure, and they were assigned as 0.15, 0.20, and 0.30 for the test program. All the samples were confined by the cell pressure of 65 kPa. The text program is shown in Table 1.

Table 1. Test program in terms of stress levels and loading types.

CSR	Double amp. axial str. (kPa)	Type 1	Type 2	Type 3	Type 4	Type 5	Type 6
0.15	20	X	X	X	X	X	X
0.20	26	X	X	X	X	X	X
0.30	39	X	X	X	X	X	X

As seen in the table, a total of 18 tests was programmed to see the effect of the stress level and loading characteristics. It is customary to use harmonic loading in the liquefaction studies; however, irregular loading shapes were used in this study to investigate the differences. Six different patterns were used to axially load the samples, and the double amplitude of axial stresses was kept constant at each CSRs. The loading shapes at CSR = 0.20 level are presented in Figure 3.

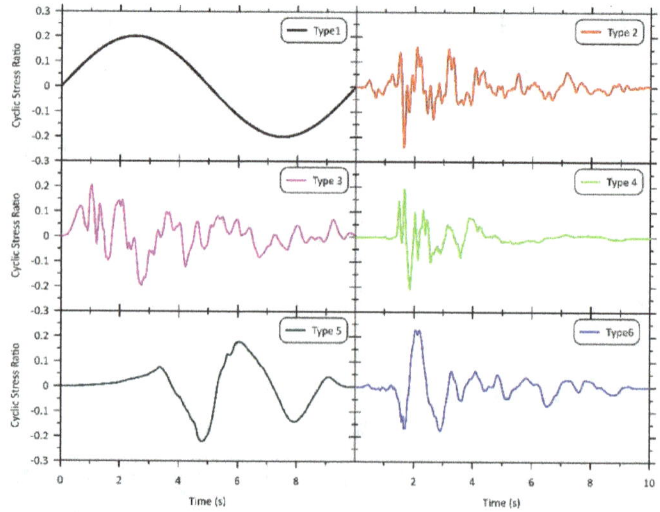

Figure 3. Example loading shapes at the level of CSR = 0.20.

2.3 Excess pore pressure generation models

Since the tests were assessed as stress controlled, the stress-based models were chosen to measure their performance with the laboratory results. Four different equations suggested by the models are presented in Table 2.

Table 2. Models and associated equations used to compare the test results.

Models	Equation	Notes
Seed et al. [13]	$r_u = \frac{1}{2} + \frac{1}{\pi}\sin^{-1}\left[2\left(\frac{N}{N_L}\right)^{1/a} - 1\right]$	α = 0.7
Booker [14]	$r_u = \frac{2}{\pi}\sin^{-1}\left[\left(\frac{N}{N_L}\right)^{1/a}\right]$	α = 0.7
Polito et al. [15]	$r_u = \frac{2}{\pi}\sin^{-1}\left[\left(\frac{N}{N_L}\right)^{1/a}\right]$	α= 0.01166·FC+0.007397·Dr+0.01034·CSR+0.5058
Baziar [16]	$r_u = \frac{2}{\pi}\sin^{-1}\left[\left(\frac{N}{N_L}\right)^{1/2\theta}\right] + \beta\sqrt{1 - \left\{\frac{2N}{N_L} - 1\right\}^2}$	$0.6 \le \theta \le 0.8$ $0.0 \le \beta \le 0.25$

In the table, N is the specific cycle that the excess pore pressure ratio to be calculated and NL is the number of cycles that liquefies the soil. α is a fitting constant, and although Seed and Booker propose as it is taken to be 0.7, Polito et al. advanced this constant with other parameters such as fine content (FC), relative density (Dr), and CSR. More complex model was offered by Baziar in 2011 with two different constants θ and β. These constants differ with the content of the silt in the sandy soil. Since clean sand samples were tested in this study, only related constants are shown in the table. The model Booker proposed gives almost identical results as Seed et al.'s model does; therefore it is not presented in the following part.

2.4 Observations

A total of 18 cyclic triaxial tests was planned to gather excess pore pressure generation at different stresses with changing excitation patterns. Using the necessary information from the test data, the excess pore pressure ratios (ru) which were the ratio of the excess pore pressure recorded during cyclic loading over the confining pressure of 65 kPa were calculated for the models mentioned in the previous section. The test data and corresponding models proposed by Seed et al. [13], Polito et al. [15], and Baziar [16] at CSR = 0.15 are presented in Figure 4.

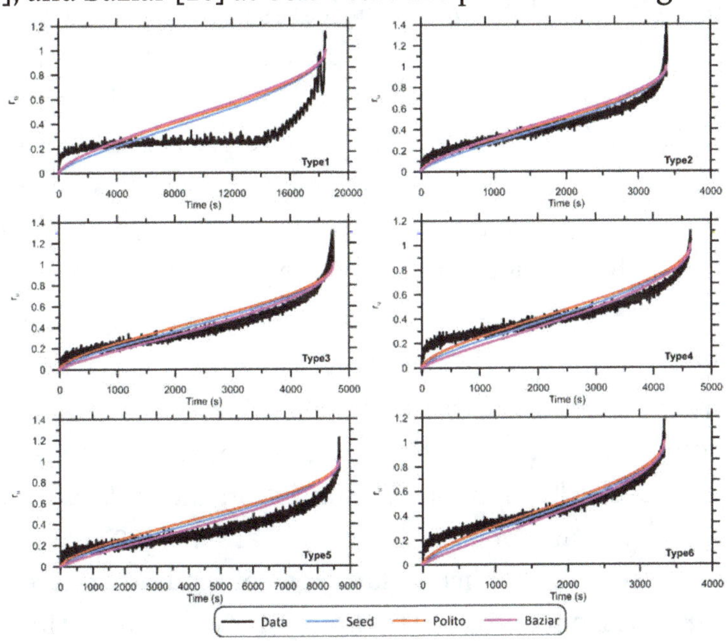

Figure 4. Excess pore pressure ratios for the test data and models at CSR = 0.15.

The first thing to point out in the figure is that the liquefaction happens in various times for different types of loading, although the double amplitude axial stresses are the same for all the six tests. The longest time to liquefy the clean sand sample happened to be almost 18,500 seconds for the sinusoidal type of loading, whereas this duration came out to be in an interval of 3500–8600 seconds for irregular type of loading. As seen in the upper left side of the figure, the recorded pore pressure buildup differs a lot with regard to other types. An initial increase was recorded in a short time, and it fluctuated for almost 10,000 seconds, and then the accumulation of the excess pressure started to increase again, and then liquefaction finally occurred after 5000 seconds for Type 1. Similar trends were observed for the rest of the tests for shorter periods.

Regarding the models suggested in the literature, they do not behave much different at this stress level. Although they did not do a good estimation to represent the data gathered from the sinusoidal loading, other pore pressure ratios along time were predicted alright by the models, being slightly on the conservative side. It should be noted that all three models missed the initial increases in pore pressure regardless of type of loading. Lastly, the pore pressure observed from the type five loading acts more like a harmonic loading than irregular type of loading, and models overestimated the pore pressure generation after 2000 seconds until the liquefaction developed. Overall, some model has better presentation of the observed data at certain type of loading, and some does at others. Therefore, no model perfectly matches with the test data.

Moving on to the higher stress level, Figure 5 shows the recorded excess pore pressure ratios at CSR = 0.20 for six different tests and corresponding estimations by the literature models. Compared to the lower stress level, the liquefaction initiation did not take that long under cyclic stress ratio of 0.20. It occurred around 130–420 seconds for soil to lose its stability for irregular types of loading, whereas it took almost 2.5–8 times longer for soil to liquefy under harmonic loading. The literature models ideally followed the pore pressure record gathered from the three tests under loading of Type 3, Type 5, and Type 6; Baziar, being the best, did acceptable predictions for other two tests (Type 2 and Type 4 loadings).

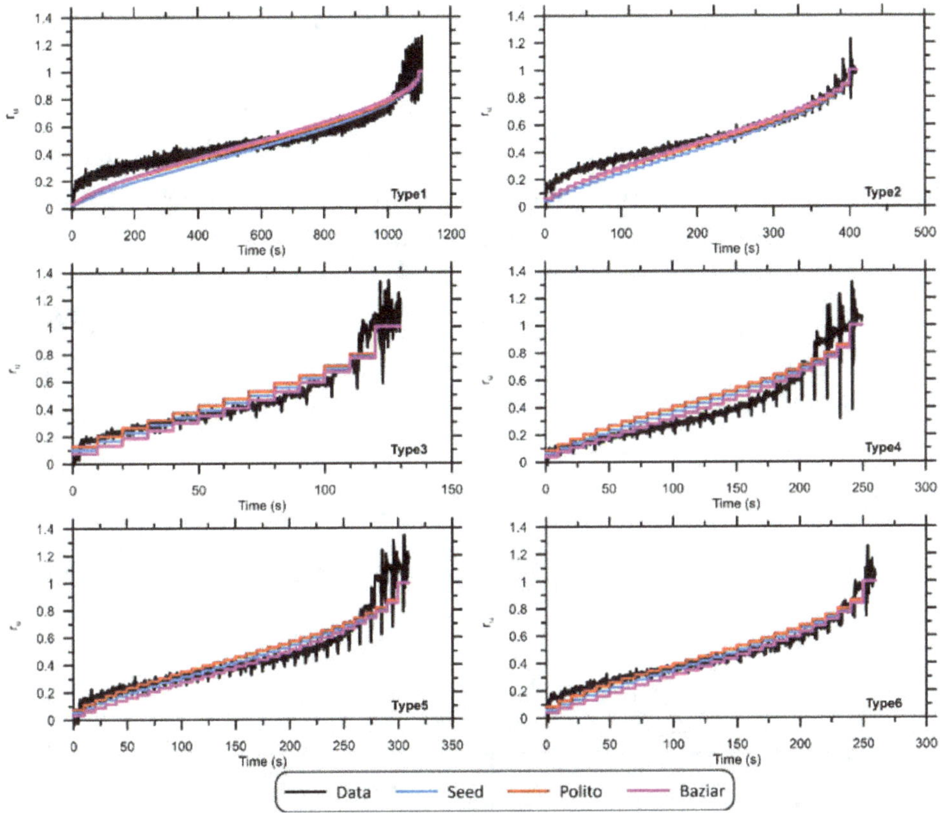

Figure 5. Excess pore pressure ratios for the test data and models at CSR = 0.20.

As the axial stress increases, the pore pressure buildup behavior under sinusoidal loading converges to the one loaded by the irregular type. There is still a mismatch at the initial, but after 400 seconds, all three models presented the data alright for the Type 1 loading. It is obvious that most of the time the model proposed by Baziar [16] is the closest to the recorded data, and Polito's model [15] gives the upper limit out of the three models, whereas the other model stayed in between two.

Final set of tests were carried out at a high stress (CSR = 0.30) and recorded excess pore pressure ratios over time, and corresponding literature predictions are presented in Figure 6. It is obvious that the increasing stress level causes soil to liquefy at low number of cycles. For all the types of loading, it did take only 1–10 cycles for soil to lose its rigidity meaning not more than 100 seconds. As seen in the figure, all of the literature models predicted the pore pressure generation almost perfectly for the test done using Type 2 loading. Regarding the others, the data are underestimated by all the models especially the test data observed from the harmonic loading. It is hard to estimate the performance of the models to represent the data under Type 5 loading since it took only 1 cycle (10 seconds) to liquefy the soil.

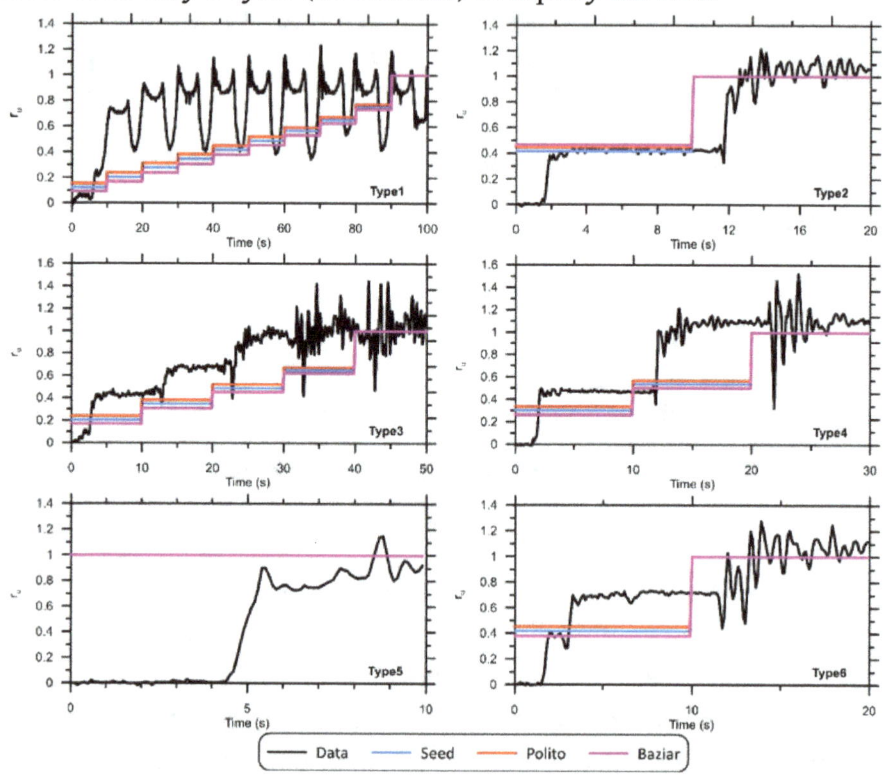

Figure 6. Excess pore pressure ratios for the test data and models at CSR = 0.30.

There is one question that can arise to understand the behavior of the liquefaction occurred when harmonic loading was applied: Did the soil liquefy at 100 seconds or earlier like 30 seconds where the first time ru value reached 1.0? During the specific test, the deformation of the soil was carefully observed with the naked eye along with the computer screen. Although ru came to 1.0 around 30 seconds, the soil was not totally unstable and still was able to resist more loads. Therefore, the total stability lost was waited to terminate the test, and it lasted almost 100 seconds.

2.5 Discussion

Out of 18 cyclic triaxial tests to understand the soil liquefaction susceptibility, it is observed that the shape of the loading or in other words frequency content of the excitation plays an important role to cause instability. Earthquakes are never harmonic types of excitations, and using sinusoidal loading in order to understand the soils dynamic behavior may mislead the interpretation. Figure 7 explains how many cycles are needed to liquefy soil sample for different types of loading at a double amplitude of 26 kPa stress, just to give an example.

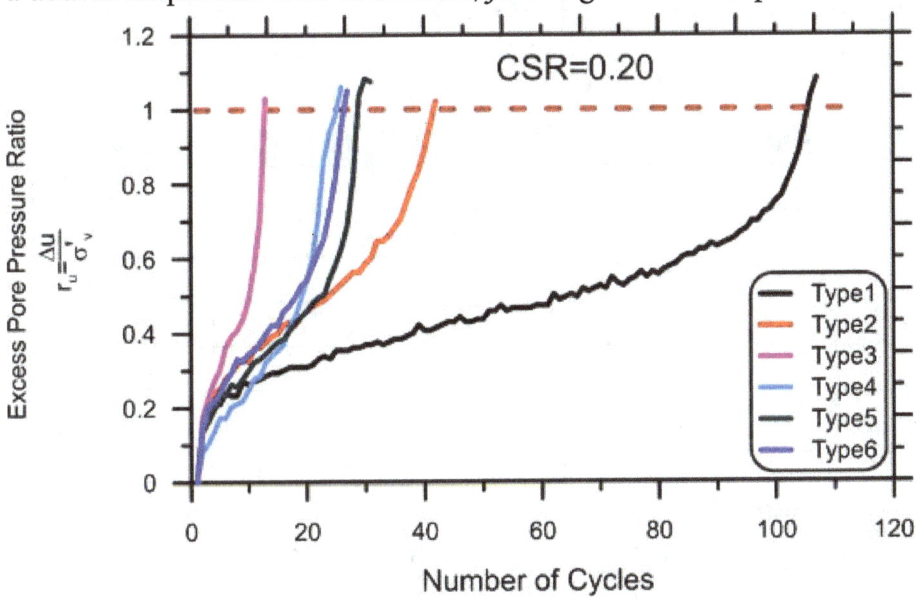

Figure 7. The effect of loading on the change of number of cycles to liquefy soil.

One clear point can be made from the figure that the variety of loading and its dominant frequency are effective on the number of cycles that would initiate the liquefaction. Although the excess pore pressure buildups are similar for the first 5 cycles, Type 1 with the harmonic shape diverges from the rest of the group, and the number of cycles needed for liquefaction occurs to be at least 2.5 times later than the others. With regard to the irregular type of loadings, they even act differently, and the number of cycles to liquefy the soil varies between 10 and 40 which can be considered a wide interval. Thus, the potential of liquefaction triggering should be studied with different types of loading when needed.

The essence of the results in terms of number of cycles to liquefy soil samples is shown in Tables 3 and 4.

Table 3. Number of cycles to initiate liquefaction for different types of loading.

CSR	Type 1	Type 2	Type 3	Type 4	Type 5	Type 6
0.15	1848	340	474	466	867	335
0.20	111	41	13	25	31	26
0.30	10	2	5	3	1	2

Table 4. Soil profiles.

Boring	Depth (m)	V_s (m/s)	$V_{s,30}$ (m/s)	FC (%)
B1	20	150–280	220	10–85
B2	20	180–190	185	5–42
B3	20	165–290	230	27–35
B4	20	145–195	175	8–32
B5	20	165–175	170	18–38

It should be noted that the variation is a lot for different cyclic stress ratios. For example, Type 3 and Type 4 loading patters need similar number of cycles to liquefy the soil at CSR = 0.15, whereas it changes for increasing CSRs. Harmonic loading always takes more time/cycle to generate pore pressure than others, and every irregular loading type has its varying frequency-dependent characteristics at increasing stresses. Therefore, the excess pore pressure generation is not only affected by the frequency content of the loading alone, but also the stress levels along with it play important role in estimating liquefaction triggering.

As a summary, three different stress levels at varying loading shapes were used to run dynamic triaxial tests in order to determine the generation of excess pore pressures under cyclic excitations. The desired relative density and confining pressure for all 18 tests were constant as 40% and 65 kPa, respectively. There are a few clear points that are worth to emphasize:

- The pattern of the loading (harmonic vs. inharmonic) is an effective parameter that influences the duration and the number of cycles to liquefy same soil samples.
- As the stress level increases, the number of cycles for soil to lose its stability decreases. For example, the needed time to liquefy soil at CSR = 0.15 changes to between 3500 and 18,500 seconds, and these values drop to 130–1100 seconds at CSR of 0.20. Even a small difference in the applied stress amplitude may cause a drastic drop in the number of cycles.
- There are many stress-based models to estimate the pore pressure buildup in the literature, and three of them were used here to see their performance to estimate the laboratory data. Models have better representation of the pore pressure generation at the low to medium stress levels for irregular loading patterns, and at high stress levels the capability to estimate the data is not that well. Among the models, the model Baziar et al. [16] suggests relatively does better jobs, and Polito et al.'s [15] model is being on the safer side all the time. However, the performance of all three to predict the recorded data does not differ that much.
- The dynamic triaxial tests take some time to set the sand sample appropriately and evaluate the data to understand the behavior, but other levels of relative densities and confining pressures should be tested to have better understanding of the pore pressure generation concept.
- The latest point is that the effect of fine content should also be investigated, and if the opportunities (budget, facility, etc.) are possible, undisturbed samples should directly be taken out of the site to see if the real in situ samples are represented well by the laboratory approaches in order to evaluate the response of the pore pressure buildup.

3. NONLINEAR SITE RESPONSE OF LIQUEFIED AREAS

3.1 Introduction and studied area

Japan, Chile, the USA, Italy, Iran, and Turkey are some of the most important earthquake-prone countries, and they have been exposed to devastating activities over the last decades causing damaged buildings and many fatalities. One of the most essential regions for liquefaction studies in the literature is the Duzce/Sakarya Region in Turkey which was immensely affected by Kocaeli and Adapazari (Duzce) earthquakes in 1999. Liquefaction susceptibility in structural design should be considered in zones consisting of loose sand soils because it may cause significant damages as excessive settlement and sinking of the structures. This can only be possible by the nonlinear site-specific analysis.

In this section, the nonlinear site response of liquefied areas will be investigated using the well-documented in situ data taken from the city of Sakarya which is surrounded by Istanbul, Yalova, Duzce, and Bilecik. The region lays on the extension of the North Anatolian Fault Zone [25], and it is neighboring two local faults (Sapanca and Duzce), and the map is presented in Figure 8. The area suffered a lot, and many liquefied zone were inspected during Duzce Earthquake in 1999.

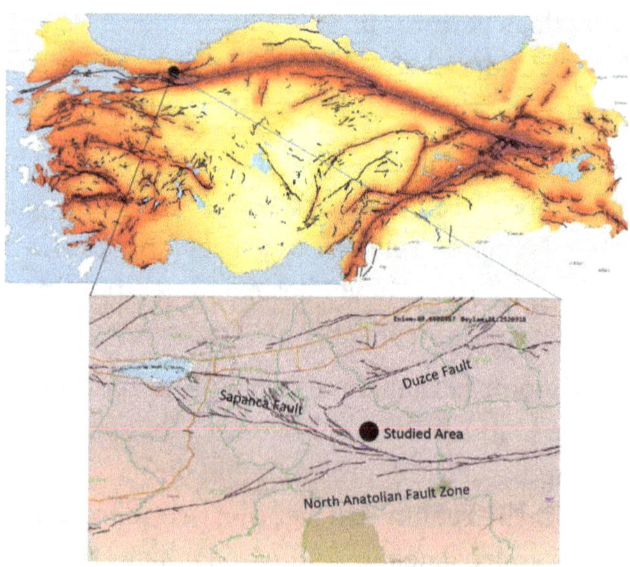

Figure 8. Studied area and the active fault lines [29].

Five different boring log data were used in the analyses to evaluate the ground response in a liquefied prone zone. The soil profiles consist of almost 90% of silty sands (SM), and some low-plasticity clays, gravelly sands, and clayey sands can be found in very shallow depths. A summary table showing the information about the logs is presented below.

All the borings are 20 m long having shear wave velocity (Vs) values changing from 145 to 290 m/s. The average Vs of upper 30 m for each profile was also determined to be able to use it in the estimation of the surface response spectra offered by the building codes, and the soil type of two borings is classified as ZE, and three of them are ZD. The fine content for the layers were obtained as well, and the variation is listed in the table. Layers have water content of 11–25%, and the unit weight of the soils samples laid in a range of 17.6–18.9 kN/m3 throughout the profiles. Some triaxial tests were run in the laboratory to evaluate the strength of the sandy soils, and 33–38 degrees of friction angle along with 0–5 kPa cohesion were detected.

The dynamic properties (modulus reduction and damping behavior of soils at varying strains) of clays and sands were calculated adopting proposed models by Darendeli [26]. During the analyses, pore pressure generation was let to build up during cyclic loading in nonlinear analyses; however it was not possible by the frequency-dependent equivalent linear simulations. With the available information of the soil index properties, the required parameters were derived to model the pore pressure generation behavior suggested by Matasovic and Vucetic [27].

3.2 Motions used in the analyses

Building codes generally split possibility of earthquake levels into some groups, and the two of them have extensive use in geotechnical and earthquake engineering which are: (1) 2% chance of being exceeded in 50 years and (2) 10% chance of being exceeded in 50 years. They offer different response spectrum envelopes for each with the help of regional earthquake maps for the seismic design. To be used in the analysis, the possible local peak ground acceleration (PGA) was determined as 0.720 g for the studied area from the seismic maps which was just published in March 2018 by the Disaster and Emergency Management Authority (AFAD) under the Ministry of Interior in Turkey. Three different earthquakes happened in last 20 years (Duzce, Kocaeli, and Van earthquakes) were chosen to produce possible regional earthquake scenarios. The details of the earthquakes used in the process are shown in Table 5.

Table 5. Reference earthquakes used in the analysis.

Earthquake	Year	Source	Station	Magnitude	R_{jb} (km)
Duzce	1999	PEER	Duzce	7.1	0
Kocaeli	1999	PEER	Izmit	7.5	1.38
Van	2011	SMDB-TR	Van	6.7	19.2

All of the earthquakes can be considered as the big magnitude earthquakes as seen in the table. Two of the events were taken from the Pacific Earthquake Engineering Research Center (PEER) [28] and one from the Strong Motion Data Base of Turkey (SMDB-TR) [29], and all of them were scaled to the local PGA accordingly. Figure 9 shows the scaled versions of the earthquakes, in other words possible earthquake schemes for the studied area for the probability of 10% chance of being exceeded in 50 years.

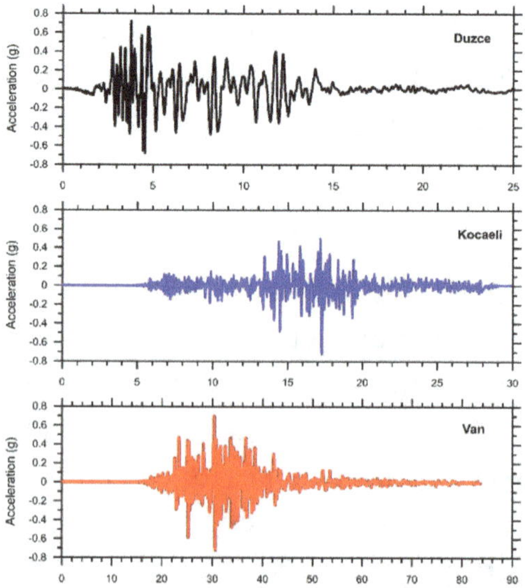

Figure 9.Scaled versions of the earthquake data.

The duration of the earthquakes differs from 25 to 85 seconds, and PGAs were set to be 0.72 g for all of them. More high-frequency content is seen from top to bottom of the figure, and these records were used as the base motions in the analyses.

3.3 Results of a specific soil profile

Site response analyses were conducted using DEEPSOIL [30] program version 6.1, and to be able to compare different approaches, five borings were excited at their base with three shakings with nonlinear and equivalent linear models for a total of 30 analyses. The results for the boring B1 shook by Duzce record, an example set, were presented in Figures 10–12 showing the NL and EL model estimations for the surface acceleration, their corresponding spectral behavior on the ground and the peak spectral accelerations, and maximum shear strains throughout the soil profile, respectively.

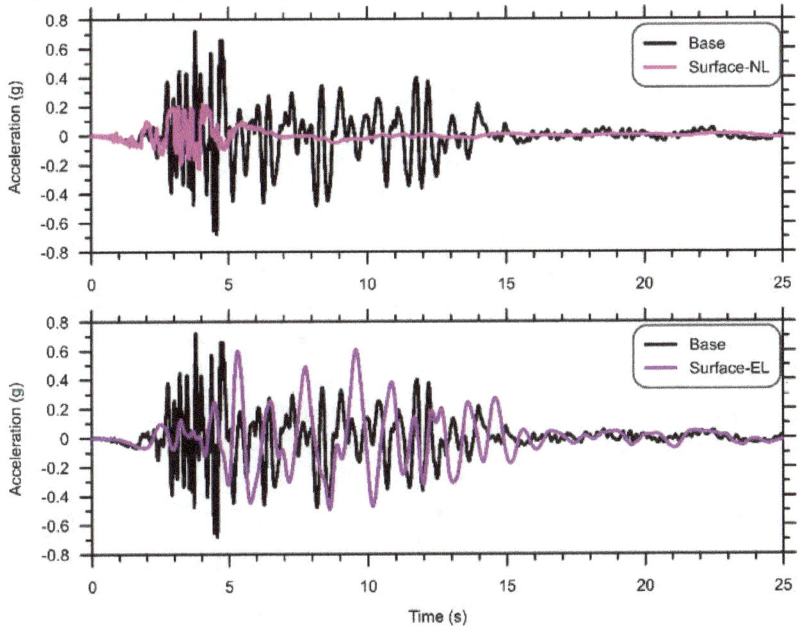

Figure 10. Base and surface acceleration observed from the nonlinear (NL) and equivalent (EL) linear models.

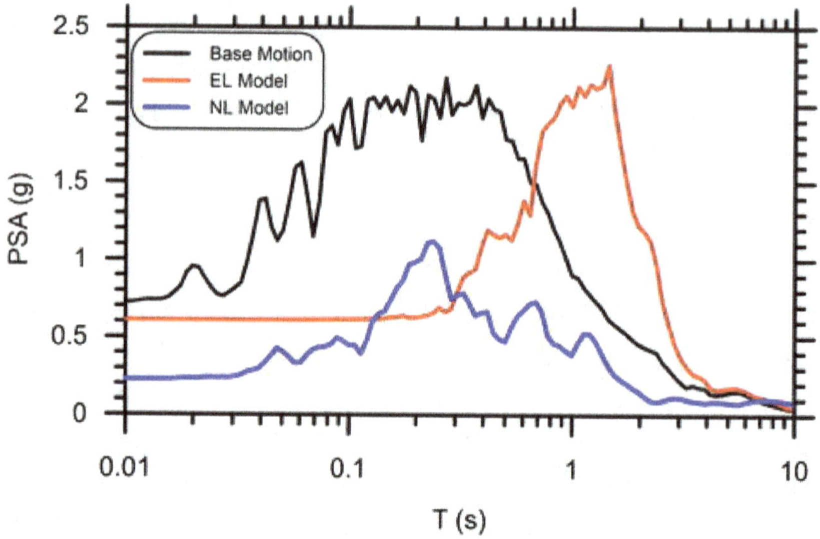

Figure 11. Spectral acceleration of the base and the surface calculated from the NL and EL models.

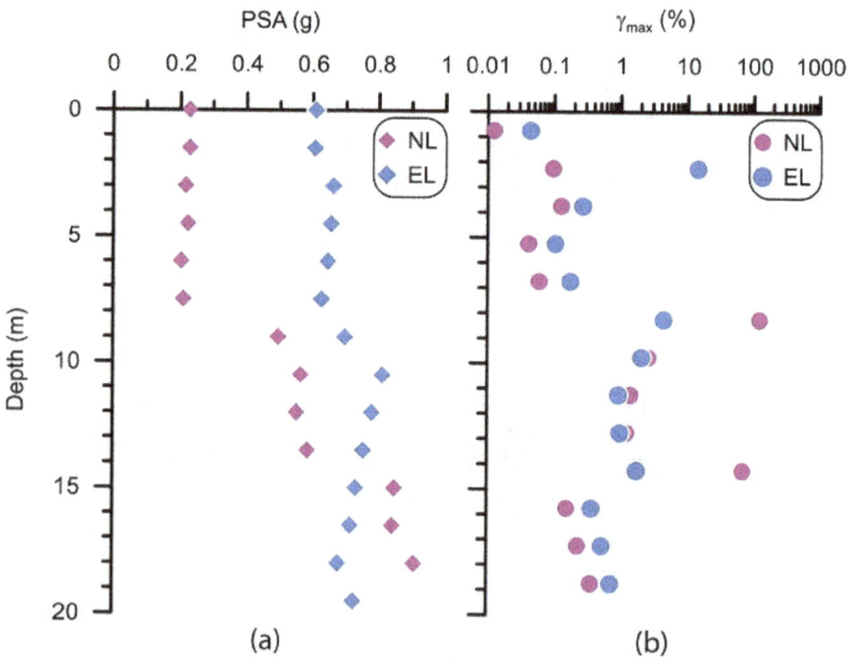

Figure 12. Peak spectral accelerations and maximum shear strains throughout the soil columns for the NL and EL models.

Regarding the surface acceleration predicted by two different approaches, there is a big deviation in the peaks and the time they occur. NL model shows that soil totally de-amplified the energy exerted by the motion, and the peak ground acceleration was noted around 0.25 g, whereas the peak-based acceleration was 0.72 g, and the main peaks happened around the similar time interval (3–5 seconds). However, an entirely different case was observed for the EL model. The prediction of the peak surface acceleration happened to be around 10 seconds extending the peak-based motion to 5–6 seconds, and not much de-amplification was noticed. In order to evaluate this behavioral diversity, the spectral acceleration at the surface was presented below.

The peak-based spectral acceleration (around 2 g) stays in a wide range of period between 0.1 and 0.8 seconds as seen in the figure. The NL model has compressed the behavior in terms of the size of peak ground and peak spectral accelerations. The model lessens the spectral acceleration 2–2.5 times at any period, whereas the EL model lengthened the peak spectral range to 0.9–1.8 seconds with almost no change in the amplitude. Another point about the EL model is that the initial flat part lasts longer than the base and NL model prediction. The last figure of the example set is the variation of the peak spectral acceleration and maximum shear strains through the soil column, and it is presented in Figure 12.

The difference between the models to estimate two parameters is concrete as observed from the figure. The peak-based acceleration was amplified a bit between 10 and 20 m and then de-amplified some for the rest of the profile by the EL model, whereas it was weakened all over the soil profile by the NL model. The soil profile was separated into three divisions, and it is supported by the related maximum shear strains at the same depths. It is obvious that two definite soil failures happened at the 8 m and 14 m according to the NL model, whereas the soil failures seem to be predicted at the depth of 2 m and 8 m by the EL model. There are still other local failures between 9 and 13 m, but the deformation level is limited compared to the ones mentioned above. It can be concluded that not only the ground response predicted by two models is quite distinct, but also the seismic actions alter the seismic behavior extremely along the soil column.

3.4 Interpretation of the analyses

The loss of soil stiffness can be caused by either insufficient bearing capacity or liquefaction triggered by the excess pore pressure generated under cyclic loading. Thus, it is beneficial to see the pore pressure generation if it was the origin of excessive deformations during seismic excitations that only NL models are capable to investigate. Excess pore pressure ratio which is a value between 0 and 1 is one vital criterion in evaluation of the liquefaction. ru of 1 means that the soil layer is under risk to liquefy for such type of loading. Figure 13 shows the excess pore pressure ratios projected by the NL models for all the logs.

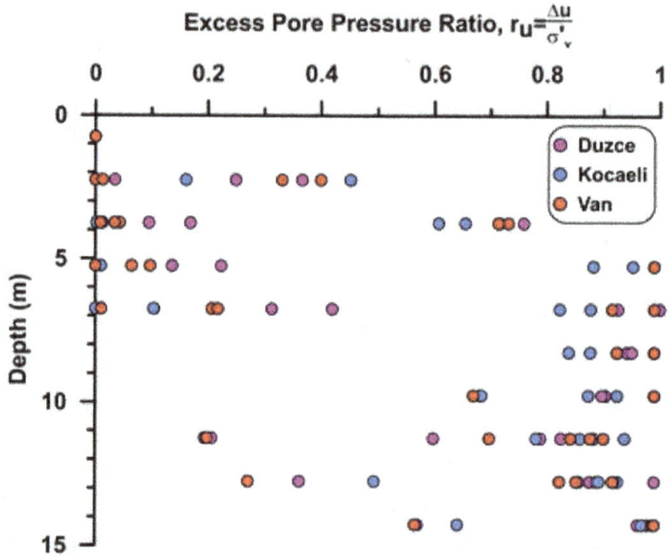

Figure 13. Excess pore pressure ratios estimated from 15 NL analysis.

One clear remark can be pointed out here that the liquefaction occurring at various depths in different logs considering all three base motions sets out the possible explanation of the unreasonable deformation indicated in the previous figure. The NL model did not anticipate liquefaction triggering for the shallow depths which include some fine content, whereas there are at least nine points that soil layers mainly composed by silty sands are expected to lose their strength due to liquefaction for such sequence of events. The ru value is greater than 0.8 for approximately 45 points meaning that even if there is no liquefaction initiated yet, soil would possibly lose its strength at these depths. Therefore, soil mitigation would be needed for possible construction in the area.

After liquefaction assessment of the soil profiles, the next step is to evaluate the site response in terms of nonlinear and equivalent linear approaches. Fifteen different analyses were conducted for each, and the peak spectral accelerations on the surface are presented in Figures 14 and 15. As mentioned before, building codes rule out the possible designs with seismic maps and some regulations. In order to see if they are good representatives of the site, corresponding response spectrum envelopes were evaluated using the new Turkish Building Code (TBC2018), Eurocode8 (EC8), and International Building Code-ASCE 7.05 (IBC), and they are shown in figures along with the ground response spectra gathered from the analyses.

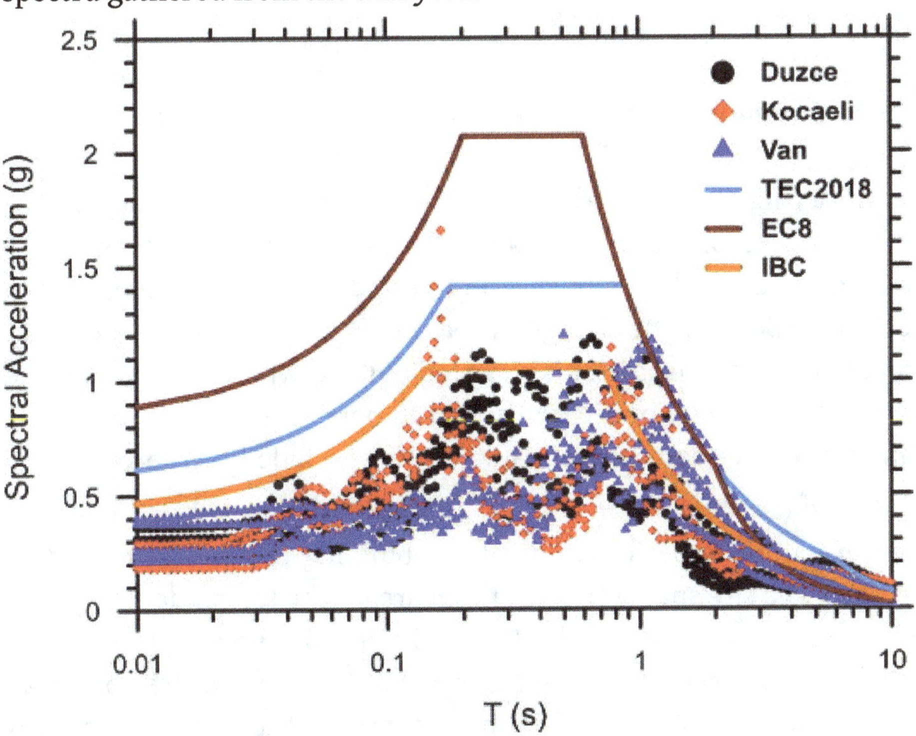

Figure 14. Ground responses evaluated from the NL analysis and building codes' design spectra.

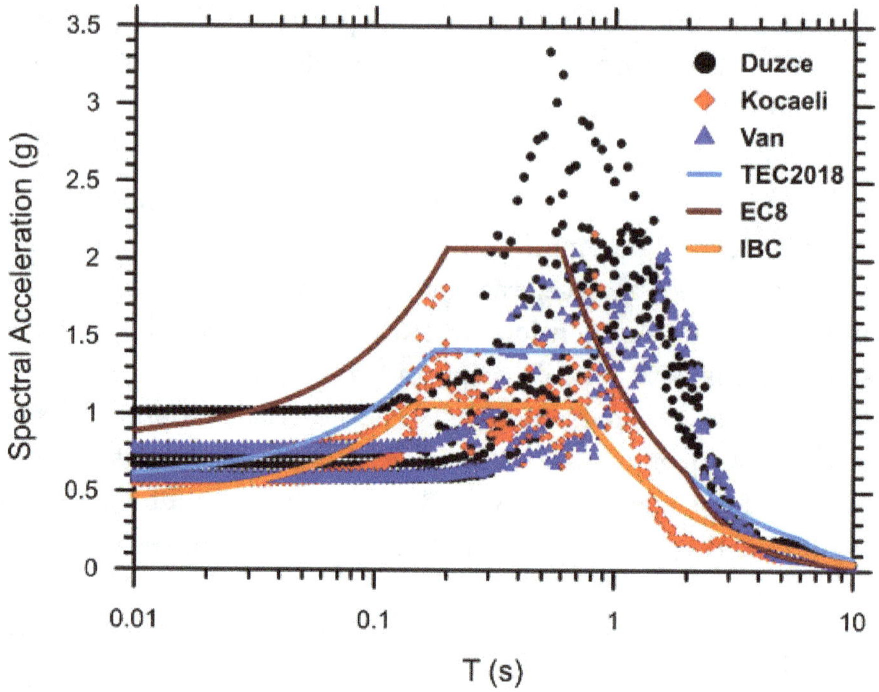

Figure 15. Ground responses evaluated from the EL analysis and building codes' design spectra.

Considering the PGAs, the variation is almost 100% from one log to another (0.20–0.40 g) in such a small area. The trend of the spectral accelerations is similar for increasing period; however the amplitudes differ a lot especially the change in peak spectral accelerations between 0.1 and 1 seconds. The lowest is reported as 0.3 g, whereas the highest is around 1.2 g with one outlier with the red dots. It is a bit off compared to the rest of the data; therefore it can be excluded to represent the general picture of the area.

In terms of the performance of the building codes with regard to expressing the site specifics, all of them are on the safer side (EC8 is the most conservative one and the IBC is the least) following the similar flow of spectral acceleration of the observed data from the nonlinear analyses. IBC's spectra would be considered as the most economical one compared to the others with missing some longer period content, and it would possibly be a better representation if the flat part of the spectra shifted on the right (by using larger corner periods). About the

TEC2018, the spectra at higher period were estimated better, but the content from small to medium periods would not be cost-effective in the seismic design. As seen in the figure, EC8 envelope is not practical at all, and such design would be the most expensive one among three building codes.

Equivalent linear analyses are easy to perform with less parameter needed to constitute a site and widely used in the practice and literature. However, one should be cautious to use this approach in such cases. Since this frequency-based analytical approach is not suited for liquefaction analyses, the results may canalize the engineer to an inappropriate design. Although the details of the results obtained from the equivalent linear analyses will not be covered here, it is important to call attention to the differences. The spectral acceleration of 15 analyses is presented in Figure 15 with their corresponding envelopes suggested by building codes.

As depicted in the diagram, the soil exhibited amplification of the base motion, resulting in an elongation of the motion duration at higher periods for the given earthquake scenarios. Neither the peak spectral acceleration nor the corner frequencies could be adequately conveyed by any of the architectural codes. It is important to note that the issue lies not with the building codes' performance, but rather with the deficient analytical approach of the equivalent linear analysis in accurately modeling the creation of excess pore pressure and the damping behavior of the soil. Hence, it is imperative to consider the non-linear response of individual sites in liquefied areas throughout the design process. Furthermore, the building regulations should include a more precise explanation of the notion of liquefaction susceptibility.

4. CONCISE SUMMARY

Formation of pore pressure The phenomena of liquefaction triggering under cyclic loading is significant, and the frequency content of the loading greatly influences the number of cycles and/or duration required to liquefy the soil. The study examined three distinct stress levels under different loading patterns and found that using harmonic loading to predict liquefaction onset is not a valid method for assessing the dynamic behavior of sands. In terms of the number of cycles, it required approximately 1848 cycles to transform the sand samples into a liquid state. This is at least twice as long as the time it took for Type 5 loading. Furthermore, with Type 6 loading at CSR = 0.15, this duration can increase up to 5.5 times. When comparing the liquefaction trigger times at a CSR level of 0.20, it is observed that the shortest time is 130 seconds, but for harmonic loading it lasts 1110 seconds, which is over nine times longer. Even within the category of irregular loading types, the discrepancy was determined to be 2.5–3 times in terms of the number of cycles required for soil to lose its stability under varying stress levels. Hence, the relationship between the frequency content and the amount of cyclic stress ratio (CSR) is crucial in understanding the formation of excess pore pressure. Regarding the stress-based models indicated in the literature for estimating recorded data in the laboratory, it was observed that these models were not very effective in predicting the data. However, it was noted that the most accurate predictions were obtained at the medium stress level (CSR = 0.2). In addition, the model suggested by Baziar [16] shown superior performance in presenting the test results compared to the other two models. Although the process of collecting sufficient information to simulate ground response in liquefied zones can be complex, nonlinear site response is the most suitable method for modeling seismic behavior. The in situ data was collected from the Duzce area, which experienced liquefaction during the 1999 Adapazari earthquake. Nonlinear analyses were conducted to replicate the liquefaction triggering at 45 distinct places at various depths, considering alternative earthquake scenarios. Due to its inability to induce liquefaction triggering, the equivalent linear technique modifies the propagation of shear waves, hence impacting the spectral accelerations and maximum shear stresses across the soil layers. It was observed that the PSA (Peak Shear Acceleration) was calculated to be three times greater at shallow depths, along with matching shear strains as an illustrative example. Therefore, it is not advisable to utilize it for assessing ground response in locations prone to liquefaction in order to avoid misinterpreting the dynamic behavior of soils. Among the three building codes used to project the site reaction, the International Building Code is the most suitable for matching the results of nonlinear analysis. In conclusion, the frequency content has a significant influence on the buildup of excess pore pressure. It is important to consider how this impact varies at different stress levels during dynamic triaxial testing. Given the complexity of nonlinear site-specific analyses, it is advisable for designers to conduct nonlinear analyses to accurately model soil behavior and ensure the safety of superstructures. Additionally, building regulations should be enhanced to address liquefaction susceptibility.

AT A GLANCE

Recent investigations on liquefaction initiation have been widely published in the literature, with a focus on dynamic triaxial tests under harmonic stress and site response of liquefied zones. Although sandy-like or clayey-like behavior is significant for evaluating liquefaction susceptibility, other cyclic loading features, such as frequency content and stress level, must also be considered. Furthermore, 1-D ground reaction analyses are used to investigate behavioral transmission down the soil column in liquefiable locations. The study focuses on two major features of liquefaction. The first section focuses on estimating pore pressure generation under irregular excitations, while the second portion evaluates the effectiveness of building codes in anticipating nonlinear site response in liquefaction prone areas. The laboratory results reveal that frequency content has a significant influence on liquefaction at different stress levels. Furthermore, there are differences across literature models for estimating pore pressure generation under various loading conditions. Regarding the site reaction, it was stated that the equivalent linear technique is incapable of anticipating the seismic behavior of the soil column; so, nonlinear ground response must be performed instead, and the IBC is the most effective in matching the nonlinear analysis results.

REFERENCES

1. Seed HB, Idriss IM. Analysis of soil liquefaction: Niigata earthquake. Journal of the Soil Mechanics and Foundations Division. 1967;94(SM3):83-108
2. Seed HB. Landslides during earthquakes due to soil liquefaction. Journal of the Soil Mechanics and Foundations Division. 1968;93(SM5):1053-1122
3. Seed HB, Peacock WH. Test procedures for measuring soil liquefaction characteristics. Journal of the Soil Mechanics and Foundations Division. 97:1099-1119
4. Finn WDL, Lee KW, Martin GR. An effective stress model for liquefaction. Journal of the Geotechnical Engineering Division. 1977;103(GT6):517-533
5. Finn WDL, Ledbetter RH, Wu G. Liquefaction in Silty Soils: Design and Analysis, Ground Failures under Seismic Conditions, Geotechnical Special Publication 44. New York: ASCE; 1994. pp. 51-76
6. Castro G, Poulos SJ. Factors affecting liquefaction and cyclic mobility. Journal of the Geotechnical Engineering Division. 1977;106(GT6):501-506
7. Castro G. On the behavior of soils during earthquakes—Liquefaction. In: Proceedings, NSF/EPRI Workshop on Dynamic Soil Properties and Site Characterization; EPRI NP-7337; Vol. 2; Palo Alto, California: Electric Power Research Institute; 1991. pp. 1-36
8. Youd TL, Hoose SN. Liquefaction susceptibility and geologic setting. In: Proceedings, 6th World Conference on Earthquake Engineering; New Delhi; Vol. 3, 1977. pp. 2189-2194
9. Youd TL. Geologic effects-liquefaction and associated ground failure. In: Proceedings, Geologic and Hydrologic Hazards Training Program. Open File Report 84-760, U.S. Geological Survey; Menlo Park, California; 1984. pp. 210-232

10. Seed HB, Tokimatsu K, Harder LF, Chung RM. Influence of SPT procedures in soil liquefaction resistance evaluations. Journal of Geotechnical Engineering. 1985;Ill(12):1425-1445

11. Kwan WS, Sideras S, El Mohtar CS, Kramer S. Pore pressure generation under different transient loading histories. In: Tenth U.S. National Conference on Earthquake Engineering Frontiers of Earthquake Engineering Anchorage; Alaska; 2014

12. Lee KL, Albaisa A. Earthquake induced settlements in saturated sands. Journal of the Soil Mechanics and Foundations Division. 1974;100(GT4)

13. Seed HB, Idriss IM, Makdisi F, Banerjee N. Representation of Irregular Stress Time Histories by Equivalent Uniform Stress Series in Liquefaction Analyses, EERC 75-29. Earthquake Engineering Research Center, University of California: Berkeley; 1975

14. Booker JR, Rahman MS, Seed HB. A computer program for the analysis of pore pressure generation and dissipation during cyclic or earthquake loading. Berkeley: Earthquake Engineering Center, University of California; 1976

15. Polito CP, Green RA, Lee J. Pore pressure generation models for sands and silty soils subjected to cyclic loading. Journal of Geotechnical and Geoenviromental Engineering. 2008;134(10):1490-1500

16. Baziar MH, Shahnazari H, Sharafi H. A laboratory study on the pore pressure generation model for Firouzkooh silty sands using hollow torsional test. International Journal of Civil Engineering. 2011;9(2):126-134

17. Byrne PM. A model for predicting liquefaction induced displacement. In: Proceedings, 2nd International Conference on Recent Advances in Geotechnical Earthquake Engineering and Soil Dynamics; St. Louis, Missouri; Vol. 2. 1991. pp. 1027-1035

18. Dobry R, Ladd RS. Discussion to soil liquefaction and cyclic mobility evaluation for level ground during earthquakes, by H.B. Seed and Liquefaction potential: Science versus practice, by R.B. Peck. Journal of the Geotechnical Engineering Division. 1980;106(GT6):720-724

19. Dobry R, Ladd RS, Yokel FY, Chung RM, Powell D. Prediction of Pore Water Pressure Buildup and Liquefaction of Sands during Earthquakes by the Cyclic Strain Method, NBS Building Science Series 138. Gaithersburg, Maryland: National Bureau of Standards; 1982. 150 p

20. Dobry R. Mohamad R. Dakoulas P. Gazetas G. Liquefaction evaluation of earth dams—A new approach. In: Proceedings, 8th World Conference on Earthquake Engineering; Vol. 3; 1984. pp. 333-340

21. Vucetic M, Dobry R. Effect of soil plasticity on cyclic response. Journal of Geotechnical Engineering. 1991;117(1):87-107

22. Carlton B. An improved description of the seismic response of sites with high plasticity soils, organic clays, and deep soft soil deposits [thesis]. Berkeley: University of California; 2014

23. Green RA, Mitchell JK, Polito CP. An energy-based pore pressure generation model for cohesionless soils. In: Proceedings: John Booker Memorial Symposium; 16–17 November 2000; Melbourne, Australia; 2000

24. Kramer S, Hartvigsen AJ, Sideras SS, Ozener PT. Site response modeling in liquefiable soil deposit. In: 4th IASPEI/IAEE International Symposium. Effect of Surface Geology on Seismic Motion; 2011

25. Komazawa M, Morikawa H, Nakamura K, Akamatsu J, Nishimura K, Sawada S, et al. Bedrock structure in Adapazari, Turkey—A possible cause of severe damage by the 1999 Kocaeli earthquake. Soil Dynamic And Earthquake Engineering. 2002;22(9–12):829-836

26. Darendeli MB. Development of a new family of normalized modulus reduction and material damping curves [thesis]. Austin, Texas: Department of Civil, Architectural and Environmental Engineering, The University of Texas; 2001

27. Matasovic N, Vucetic M. Cyclic characterization of Liquefiable Sands. ASCE Journal of Geotechnical and Geoenvironmental Engineering. 1993;119(11):1805-1822

28. PEER. Pacific Earthquake Engineering Research Center. Available from: https://ngawest2.berkeley.edu/

29. AFAD. Disaster and Emergency Management Presidency Earthquake Department. Available from: http://kyhdata.deprem.gov.tr/2K/kyhdata_v4.php

30. Hashash YMA, Musgrove MI, Harmon JA, Groholski DR, Phillips CA, Park D. DEEPSOIL 6.1, User Manual. Urbana, IL: Board of Trustees of University of Illinois at Urbana-Champaign; 2016

07 SOILS MADE OF WEATHERED GRANITE

1. INTRODUCTION: AN OVERVIEW

Grit is an igneous rock, which means it was made by magma inside the Earth. Igneous rocks are mostly made up of quartz, feldspar, mica, and hornblende [1]. Granite has a full crystal-grain structure, a hard texture, a uniform composition, and a high compression strength (120–200 MPa). There are three main types of joints in granite because of the joints in feldspar and mica. The big difference (almost twice as big) between quartz and feldspar's expansion factors makes it easy for the surface of granite to crack when it expands and contracts. Because of this, granite would weather quickly, especially granite with a thick grain structure [2]. Besides the action of weathering, long-term geological tectonic movement and changes in the climate in a certain area also affect granite. The different levels of weathering of granite and its physical and chemical qualities make it possible to divide decomposed granitic soils into four main groups [3]. These groups are weak weathering, weathering, strong weathering, and residual.

As the civil infrastructure gets better, more and more engineering projects (like roads, railways, dams, and so on) will be built in difficult mountain places. But it's very hard to build these infrastructures in mountainous places because there aren't enough building materials. To get around the lack of building materials in mountain areas, it is very important to learn about the engineering qualities and uses of unique materials like weathered granite soil. In addition to its nonlinear stress-strain, elastic-plastic, dilatancy (shrinkage), and other features, weathered granite is unique in that it breaks down easily and is easy to weather. This process of aging controls the range of particle sizes in in situ soils. Soils' particle breakage properties are also changed by how the particle sizes are spread out. A lot of research hasn't been done on the particle size distribution of in situ weathered granite soil, but it's very important to fully understand it. This is because the particle size distribution has a big effect on the engineering properties of weathered granite soil, like how well it mixes with other materials, how permeable it is, and how strong it is when it's deformed [4, 5, 6, 7]. The weathering process, particle breakage, mechanical properties (such as compaction, bearing, strength, and shearing-dilatancy characteristics), and constitutive model of weathered granite soils from Shanxi, China, were looked into in this chapter. The findings can help us fully understand the engineering features of worn granite soils and give us ideas on how to use these soils in real-life engineering situations.

2. SPECIMEN PREPARATION AND TEST METHODS

The weathered granite soil samples used in this chapter were obtained from a site near Mount Yunzhong (shown in Figure 1), a branch of the Lu-liang Mountains, on the eastern outskirts of Xinzhou City, Shanxi Province, China. The weathered granite soil was used as a part of the subgrade materials in the Xinbao highway (Xinzhou City to Baode City). There were three colors of the weathered granite in the field, red-brown, yellow-brown, and gray, as shown in Figure 1. The in situ dry unit weight and moisture content of the weathered granite were 17.82–18.22 kN/m3 and 4.9–5.5%, respectively.

Figure 1. Photograph of weathered granite samples.

The excavation range of the filling materials used for Xinbao highway (Shanxi Province, China) was mined at a depth of less than 30 m on average under the surface of the Lu-liang Mountains. This is a general practice in producing subgrade materials for highways. Because the red-brown and the yellow-brown weathered granites were located at depths of less than 30 m, the lab tests in this chapter are focused on these two kinds of samples. The embankment is mainly filled with red-brown weathered granite, so parts of the tests in this chapter were mainly aimed at red-brown samples.

Because the physical meaning of Fukumoto's weathering model [4] is clear and easy to use, this model was used in this chapter to evaluate the weathering process of granite. After the thorough investigation on the distribution of in situ weathered granite, numerous sieving tests on a large number of granite samples obtained from typical sections A and B (shown in Figure 2) were carried out. According to Fukumoto's grading model, geological year's parameter m and geometric progression constants r of granite samples at different depths were calculated. The experimental tests involved in this chapter, including X-ray diffraction, sieving, heavy compaction [8], and large-scale triaxial test [9], were investigated to research the particle breakage characteristics of compacted weathered granite.

(a) (b)

Figure 2. Photograph of field sampling: (a) section A and (b) section B.

Figure 3 shows the composition of the rock samples in this study. Figure 3 shows that the colors of the samples became gradually brighter with an increase in the quartz content. The increase in potassium feldspar or feldspar content was the main cause of the brighter colors of the rocks. Geng et al. [10] noted that the amphibole was bluish-green, and the biotite was yellow in Mount Yunzhong. In addition, Geng et al. [10] described the main body of the granite as a coarse grain (porphyritic granite), with fine-grained granite on the edge. The three types of samples used in this chapter had quartz (mean 40.3%) and feldspar (mean 54.3%) as the dominant minerals present with smaller amounts of mica (mean 1.3%), amphibole (mean 2%), hematite (0.5%), vermiculite (1%), and unknown (1%) materials.

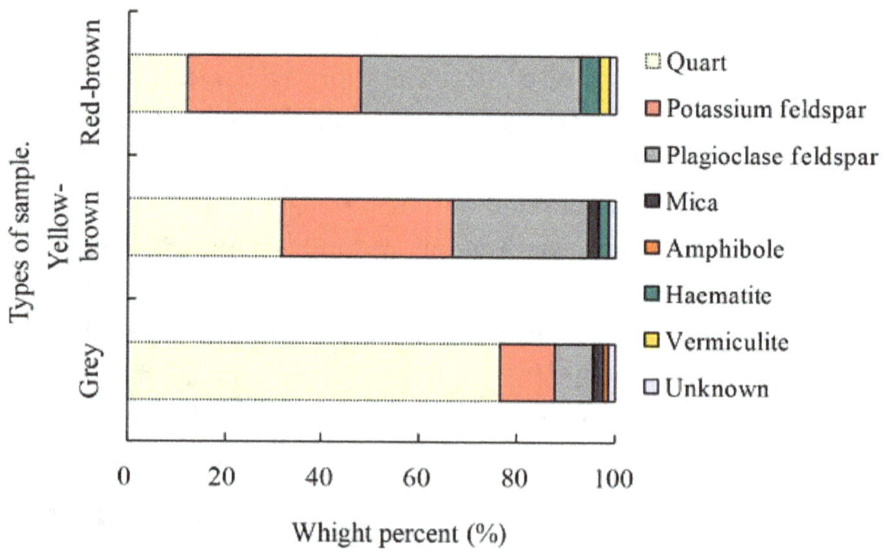

Figure 3. Mineral composition of weathered granite samples.

3. WEATHERING PROCESS

Based on the geometric fractal theories and some assumptions, Fukumoto [4] proposed a gradation equation for decomposed granite soils by using mathematical statistics method and made it possible to quantificationally describe the weathering process of granite by using a certain mathematical model to evaluate the particle breakage of weathered rock. Because Fukumoto's model is more scientific and easier to use, this model has been widely used in describing the process of rock weathering. Thus, this model was also used in this chapter to describe the weathering process of in situ weathered granite.

Figure 4 shows changes of parameters m and r for the samples at different depths of mountain profile, h.

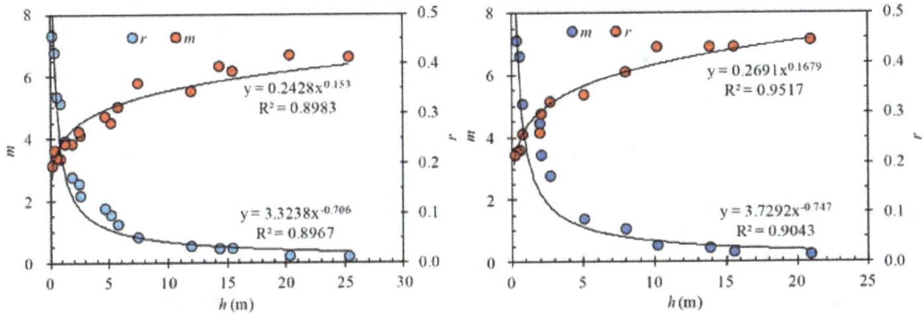

(a) (b)

Figure 4. Change of parameters m and r for samples at different depths: (a) section A and (b) section B.

The results of statistical analysis indicate an approximate power function relationship between h and m (or r), and the correlation of relationship was relatively good. The power function relationships between h and m (or r) in section A are shown as follows:

$$m_{\mathrm{A}} = 3.324h^{-0.706} \tag{1}$$

$$r_{\mathrm{A}} = 0.243h^{0.153} \tag{2}$$

where h is the depth of granite samples, which was measured from the surface of mount, m A is the geological year's parameter in section A, and r A is the geometric progression constant in section A.

The power function relationships between h and m (or r) in section B are shown as follows:

$$m_{\mathrm{B}} = 3.729h^{-0.747} \tag{3}$$

$$r_{\mathrm{B}} = 0.269h^{0.168} \tag{4}$$

where m B is the geological year's parameter in section B and r B is the geometric progression constant in section B.

Actually, the geological year's parameter, m, reflects the degree of geological actions (including physical actions and chemical actions) [11]. The deeper the granite is buried, the less the geological action will affect it. The above results indicate that m was close to zero at a depth of 20–25 m under the surface of the mount in sections A and B, which indicate that the physical and chemical actions on the granite in this research can be negligible at this depth. The geometric progression constant, r, represents the integrality of rock. The r of section A and section B studied in this chapter was close to a certain value at a depth of 25 and 20 m, respectively, which indicates that the granite in Mount Yunzhong was close to the complete unweathered state at a depth of 20–25 m.

It can be found from the results that the m of section A was larger than that of section B at the same depth, indicating that the ability of section A to resist geological and climatic environment was worse than that of section B. On the other hand, the r of section A was smaller than that of section B at the same depth, which illustrates that the integrality of granite in section A was worse than that in section B [11]. Although section A and section B are under the same geological and climatic environment, the values of m (or r) are different between these two sections. The main reason for the difference was the difference in internal properties (e.g., the mineral composition of rocks) of weathered granite between these two sections. Of course, some external factors could not be excluded; for example, the probability of fracture distribution of section A is larger than that of section B.

4. PARTICLE BREAKAGE CHARACTERISTICS

Many aspects, such as particle gradation, mineral composition, blows per layer, and stress level, can influence the characteristics of particle breakage of compacted weathered granite.

Hardin [12] proposed the concept of relative breakage (B r), which was achieved from the variation in particle size distribution curves before and after loading. Relative breakage can embody the overall change in particle size distribution before and after loading and overcomes the shortcoming of using single index of sieving test results (percentage of particles passing a given sieve size or sieve size corresponding to a given percentage passing) as the evaluation index for particle breakage [13, 14, 15, 16]. In view of the above, B r has been widely used in evaluating soil particle breakage characteristics, because it can represent the basic attribute of the soil material and is more scientific and reasonable to use.

4.1 Particle gradation

Standard heavy compaction tests on four red-brown samples and four gray samples of weathered granite with different gradations under optimum moisture content conditions were investigated. Sieving tests were implemented on the samples before and after compaction, and the uniformity coefficient ($C_u = d_{60}/d_{10}$) and relative breakage (B_r) of the samples before and after the compaction test were calculated. The relationship between C_u and B_r is shown in Figure 5 [11]. The test results show that B_r decreased with the increase of C_u which indicates that as the C_u of weathered granite increased, the degree of particle breakage was reduced because of the enhancement of the intergranular locking effect. In addition, it can be seen that the rate of decrease of relative breakage of the red-brown sample was higher than that of the gray sample, reflecting that because the degree of weathering of red-brown weathered granite is higher than that of gray weathered granite, the strength of the red-brown sample was less than that of the gray sample. This shows that the sensitivity of the red-brown sample to particle heterogeneity was greater than that of the gray sample [11].

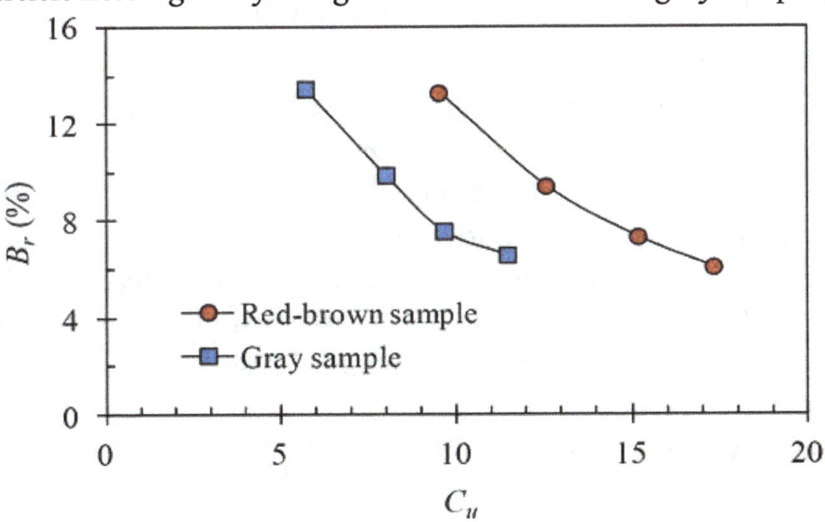

Figure 5. Br-Cu curves.

4.2 Mineral content

The red-brown, yellow-brown, and gray weathered granite samples were prepared with approximately the same particle size distribution. Standard heavy compaction tests were performed on the samples under optimum moisture content conditions [11]. The results of sieving before and after the test are shown in Figure 6.

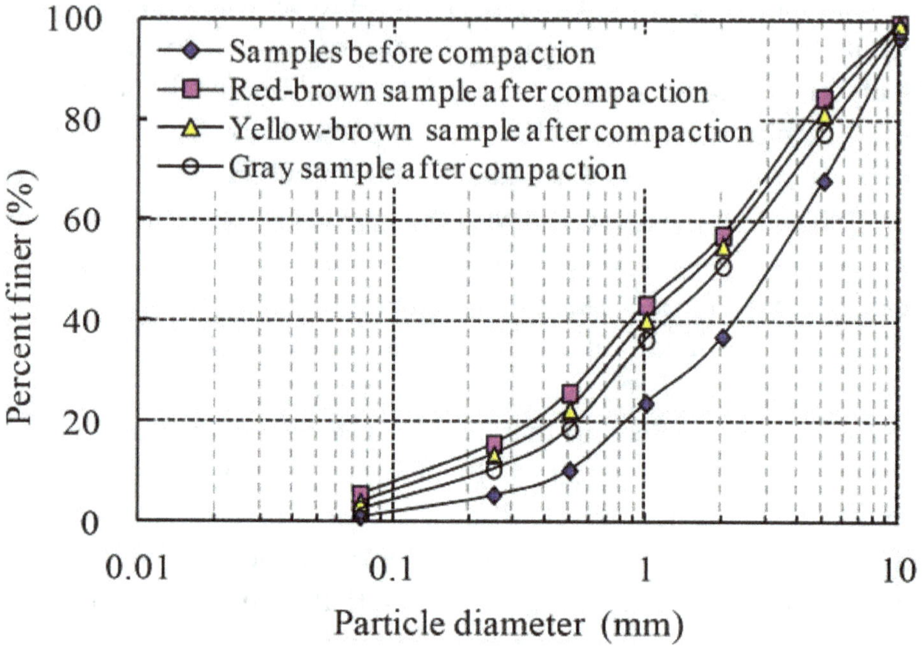

Figure 6. Sieving curves of different samples before and after the test.

The relationship between relative breakage and quartz (or feldspar) content was analyzed by calculating the relative breakage of different samples before and after the standard compaction test. Figure 7 shows the relationship between relative breakage and quartz (or feldspar) content. It can be seen from the figures that B r decreased with the increase of quartz content, while B r increased with the increase of feldspar (=plagioclase feldspar + potassium feldspar). The results show that quartz and feldspar content have an obvious effect on the particle breakage characteristics of weathered granite [11].

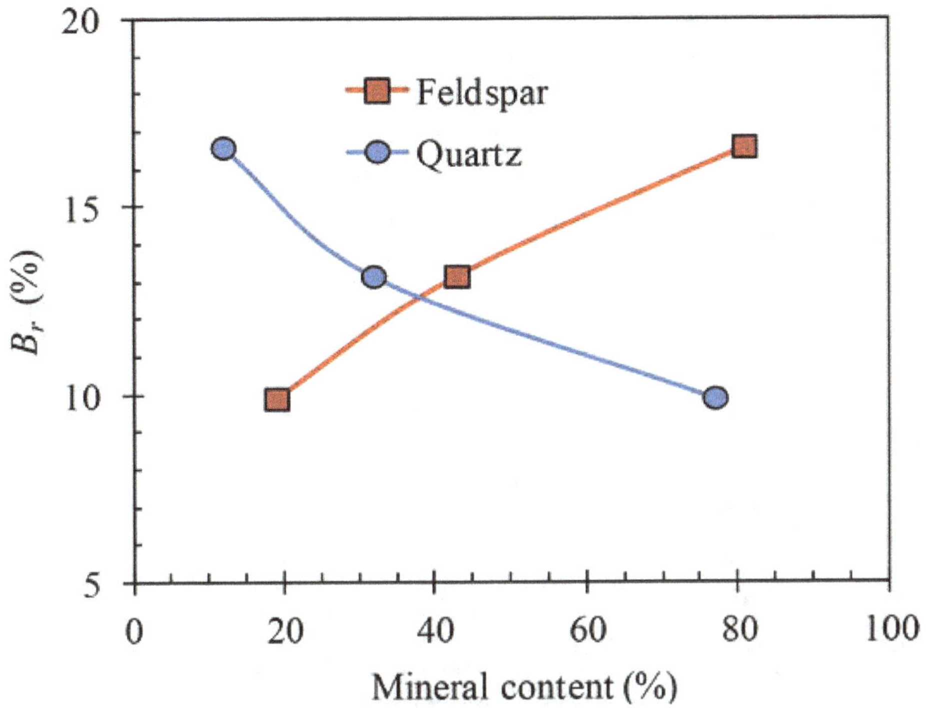

Figure 7. Relationship curve between mineral content and Br.

The main reasons that the B r of samples with high quartz content after compaction test was small are as follows: (1) the probability of breakage of samples with more quartz content is small because quartz has a high strength, (2) samples with high quartz content have strong ability to resist being weathered, and (3) there are few microcracks in samples with high quartz content. On the contrary, feldspar has little strength and is easily weathered, so the samples with higher feldspar content showed obvious particle breakage characteristics [11].

4.3 Blows per layer (compaction degree)

In order to analyze the effect of blows per layer of samples on the particle breakage properties of weathered granite, four red-brown weathered granite samples with the same initial particle size distribution were prepared, and four different heavy compaction tests were conducted, with blows per layer (BPL) of 30, 50, 75, and 98 [11]. The sieving results before and after the tests are shown in Figure 8.

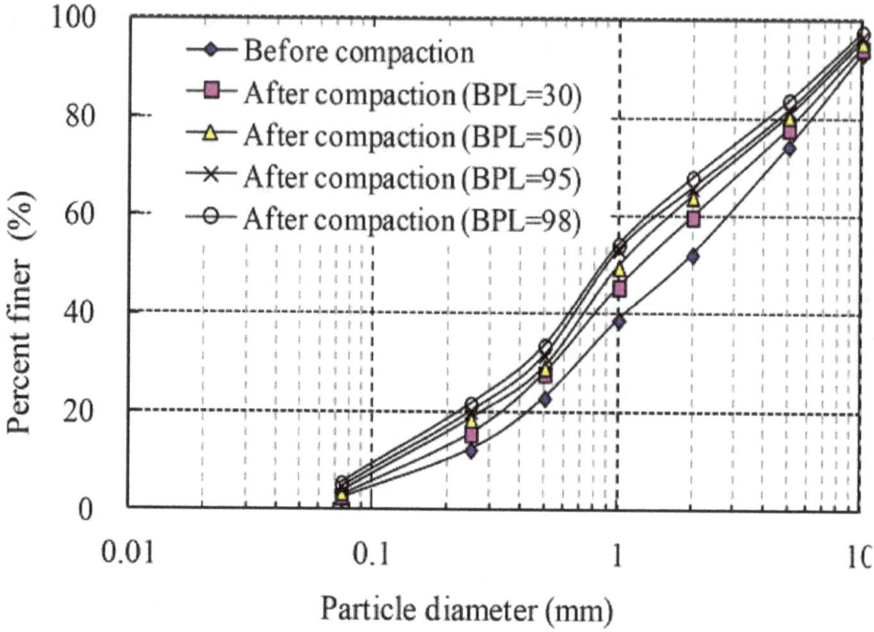

Figure 8. Sieving curves of samples under different blows per layer. BPL: blows per layer.

As shown in Figure 9, the relative breakage increased with an increase in blows per layer, but the increasing level of relative breakage decreased. Furthermore, with the further increase of blows, B r tended toward a certain limit value. Considering the engineering practice, the compaction degree of samples was analyzed, and the compaction degrees of samples and corresponding relative breakage are depicted in Figure 10. It can be concluded from this figure that there is an approximate linear growth relationship between compaction degree and B r. It can be found that for the same weathered granite fillings, relative breakage can be used to reflect the compaction performance indirectly on the basis of the relationship between compaction effect and compaction performance. Excessive compaction may lead to excessive particle breakage of soils and is not conducive to the long-term stability of subgrade [11].

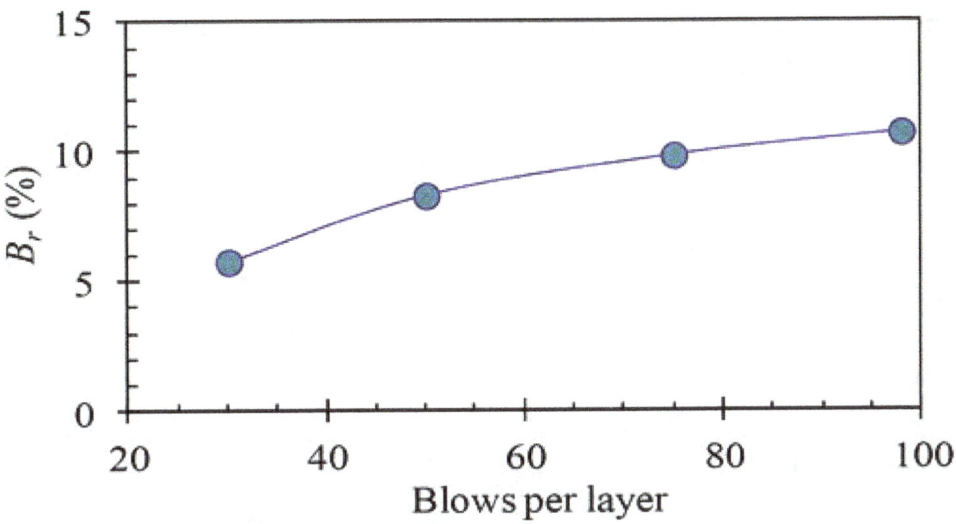

Figure 9. Relationship curve between blows per layer and Br.

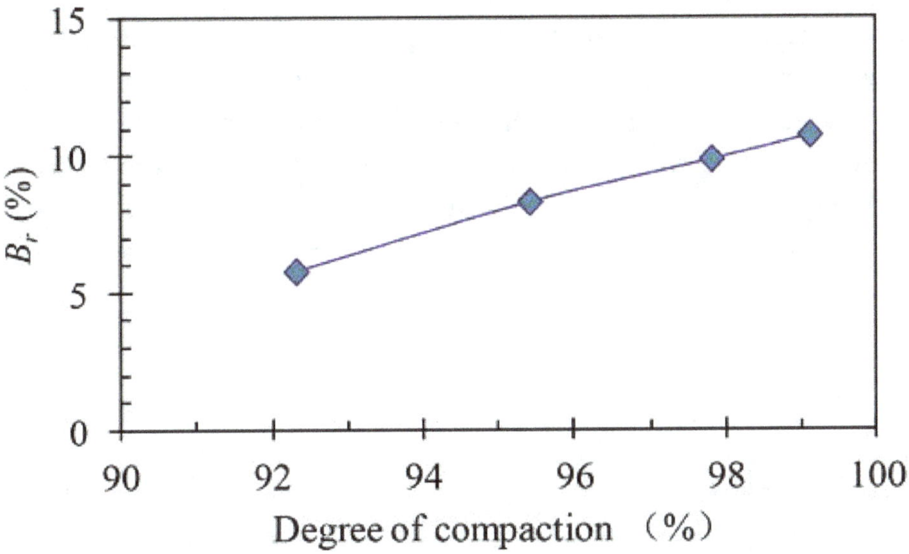

Figure 10. Relationship curve between degree of compaction and Br.

4.4 Stress level

The breakage of soil particles in a sample of rockfill under moderate stress will be quite evident [11]. For the purpose of studying the particle breakage characteristics of compacted weathered granite soil under different stress levels, particle size analysis tests were conducted on the samples before and after consolidated drained large-scale triaxial tests, and the relationship between relative breakage and confining pressure (or stress ratio, q/p) was analyzed [11].

Because specimens for the triaxial test must be prepared by the vibrating forming method, particles of the specimens may be partially crushed. In order to improve the accuracy of the experiment, two specimens for the large-scale triaxial test under the same confining pressure were artificially prepared with approximately the same particle size distribution before vibrating compaction. One of the samples was sieved after vibrating compaction and before the triaxial test. The results of the sieving test for this sample were used as the initial gradation of the other sample for the triaxial test [11].

From the stress-strain curve of samples in this study, it can be seen that when the axial strain reached 15%, the residual strength of the specimen under different confining pressures reached its constant value. Therefore, when the axial strain reached 15%, the triaxial tests were forced to stop, and then the sieving tests were performed. Figures 11 and 12, respectively, show variations in relative breakage (B r) with confining pressure (3) and stress ratio (q/p) [11].

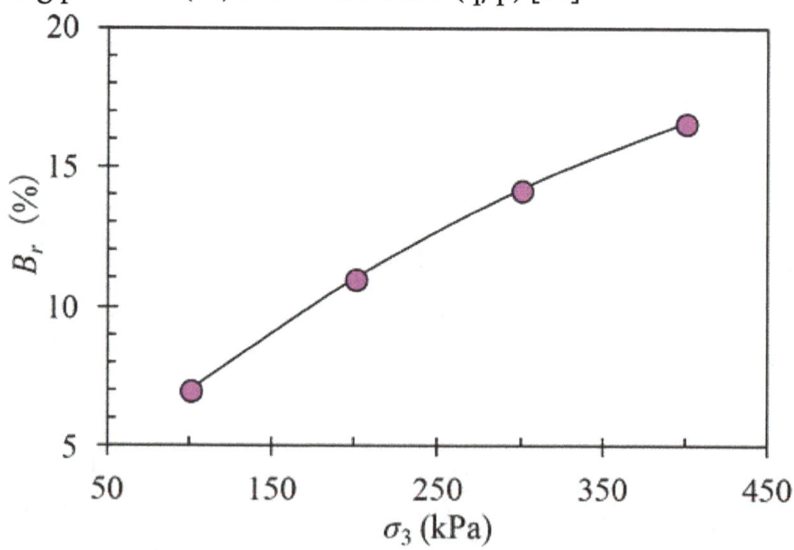

Figure 11. Br- 3 curves.

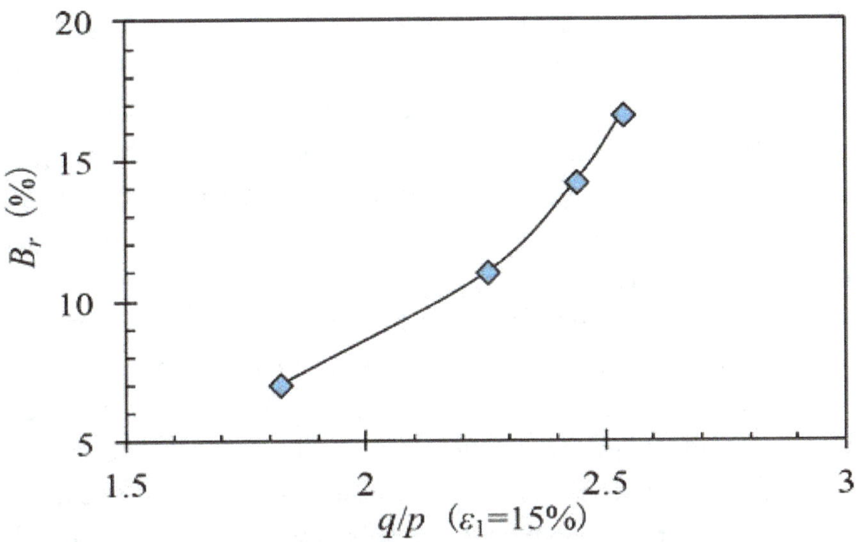

Figure 12. Br-q/p (1 = 15%) curves.

As indicated in Figure 11, at the end of the triaxial test, the relative breakage of samples increased with the increase of confining pressure, while the increasing amplitude of relative breakage decreased slightly with the increase of confining pressure. It can be considered that the higher confining pressure limited the further compaction and movement of soil particles in soil samples and resulted in producing microcracks and fracture within a single particle. The data in Figure 12 illustrate that at the end of the experiment (1 = 15%), both relative breakage B r and the amplitude of B r increased with the increase of the stress ratio q/p. These results indicate that under the same axial strain conditions, more particles were broken due to the increased strain confinement caused by higher confining pressure [11].

5. MECHANICAL BEHAVIOR

5.1 Compaction characteristics

Figure 13 shows the compaction curves for the red-brown and yellow-brown samples with different clay contents. It can be seen from Figure 13 that the peak maximum dry density (MDD) values occur most significantly when the clay contents are in the range of 7.5–10%. The peak MDD of the red-brown and the yellow-brown samples of weathered granite soil is 2.32 and 2.38 g/cm3, respectively. The results of the tests clearly show that as the clay content increases, the MDD tends to be considerably reduced after it reaches the peak MDD. Therefore, the experimental results show that the clay content of weathered granite soil has a remarkable influence on its compaction characteristics. Furthermore, at peak MDDs, the clay content of red-brown weathered granite soil is 1% larger than that of yellow-brown weathered granite soil. Because the particles of the red-brown samples were smaller than the ones in the yellow-brown samples, the yellow-brown samples mixed with clay could be easily formed into the suspended-dense structures in the process of compaction. The red-brown samples mixed with clay could be easily formed into the skeleton-dense structures. Consequently, the peak MDD of red-brown samples was smaller than that of yellow-brown samples. In addition, the higher the clay content, the larger is the optimum moisture content, and this relationship is approximately linear. These results also indicate that because the gradation of red-brown samples is finer than yellow-brown, the optimum moisture content (OMC) of red-brown samples is 0.1% smaller than yellow-brown. It is noted that the increase of OMC slowed down slightly when the clay contents of the samples were more than 8% [17].

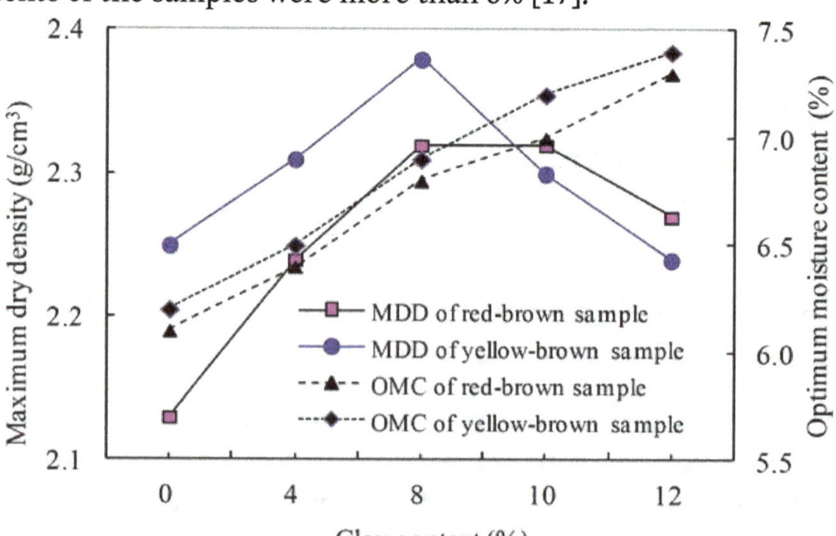

Figure 13. Compaction curves for weathered granite samples with differing clay content.

5.2 Bearing characteristics

Figure 14 represents the relationship of clay contents and CBR (California Bearing Ratio) obtained from different compaction tests, with the blows per layer of 30, 50, and 98, respectively. As shown in Figure 14, the CBR comes to a peak value when the value of blows per layer is 50 or 98. Because the compaction power is not enough to make rock material in a dense state when compacted under 30 blows, the CBRs of the samples with 30 blows do not arise to a peak value. However, with the increase of clay content ratio, the samples with 30 blows will eventually arise to a peak CBR. The experimental results show that the peak value of CBR increases with the increase of blows per layer. But the results also show that as the blows per layer increase, the clay content ratio at the point of peak CBR decreases. It is noted that the clay content ratio at the point of peak CBR with 98 blows per layer is approximately 4%, which is 4% less than the clay content ratio at the point of peak maximum dry density. The cause of the above results is mainly because that when clay content ratio exceeds a certain value (i.e., 4%), the interlocking structure of the compacted weathered-granite would be opened by the clay in the material, the internal friction angle (φ) would be decreased, and the penetration resistance subsequently would be declined. Finally, the clay content ratio at the point of peak CBR is larger than that the ratio at the point of peak maximum dry density [17].

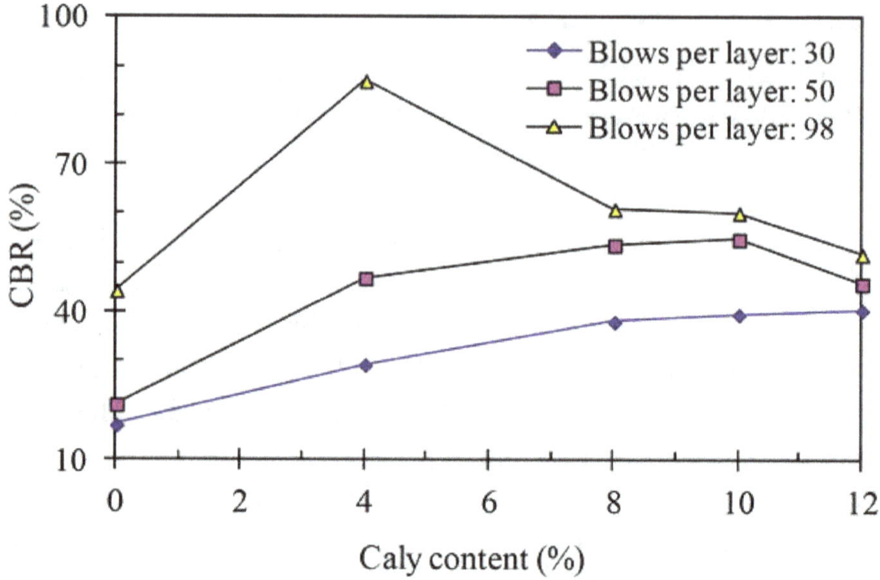

Figure 14. Change in CBR for soils with differing clay content.

5.3 Strength characteristics

Figure 15 shows the typical stress-strain relationship of the CD tests on four representative red-brown samples of pure weathered soil under four different confining pressures, where 1 is the axial stress, 3 is the confining pressure, and d 1 is the axial strain. Figure 16 shows this typical v^{-1} relationship of the CD tests, where v is the volumetric strain.

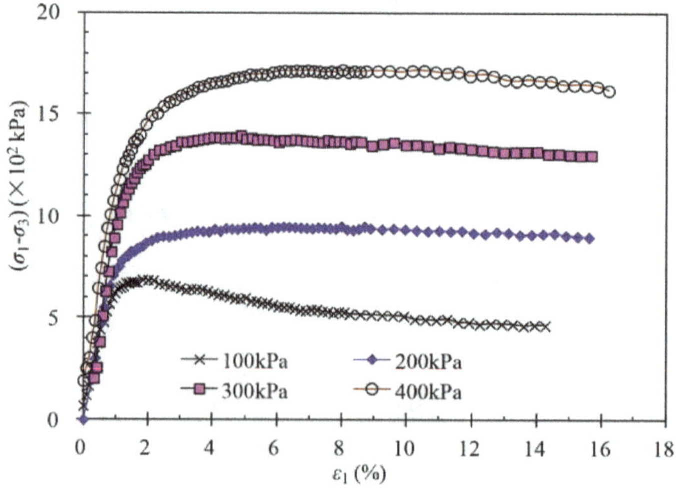

Figure 15. Stress-strain curves.

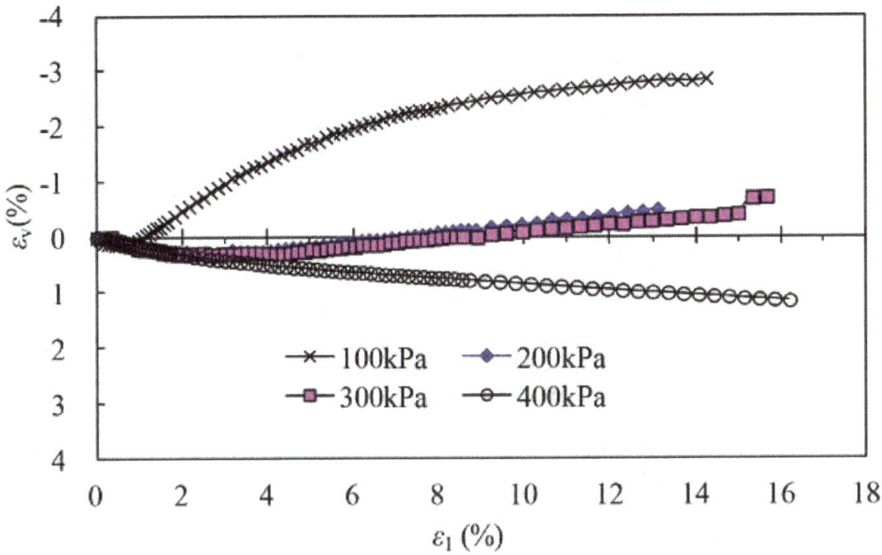

Figure 16. Curves of v- 1.

It can be found in Figure 15 that the peak deviator stress increases as the confining pressure increases. The internal particles of weathered granite soil could be overturned, stridden, and dislocated under lower confining pressure and could be crushed, filled, and compacted under higher confining pressure. It was eventually manifested that the specimens transit gradually from strain softening status (e.g., when the initial confining pressure was 100 kPa) to strain hardening status (e.g., when the confining pressure was 200 kPa) with the increase of confining pressures on the samples. At any event, there exists a critical confining pressure. The samples soften in strain when the experiment confining pressures were lower than the critical confining pressures and harden in strain when the experimental confining pressures were higher than the critical confining pressures [18]. Thereafter, it could be deduced that the critical confining pressure in this study was between 100 and 200 kPa. It can be observed from Figure 16 that the samples transit generally from shear dilation to shear shrinkage, and the volumetric strain of the transition point of the specimens increases gradually with the increase of confining pressures on the samples. In summary, as the confining pressure increases, the volumetric strain at failure changes toward volume contraction. According to the results of large-scale triaxial tests, the Mohr's circles and the strength failure envelope can be depicted in Figure 17 on the basis of the Mohr-Coulomb strength criterion. The linear Mohr's envelope of the weathered granite soil in this work can be expressed as

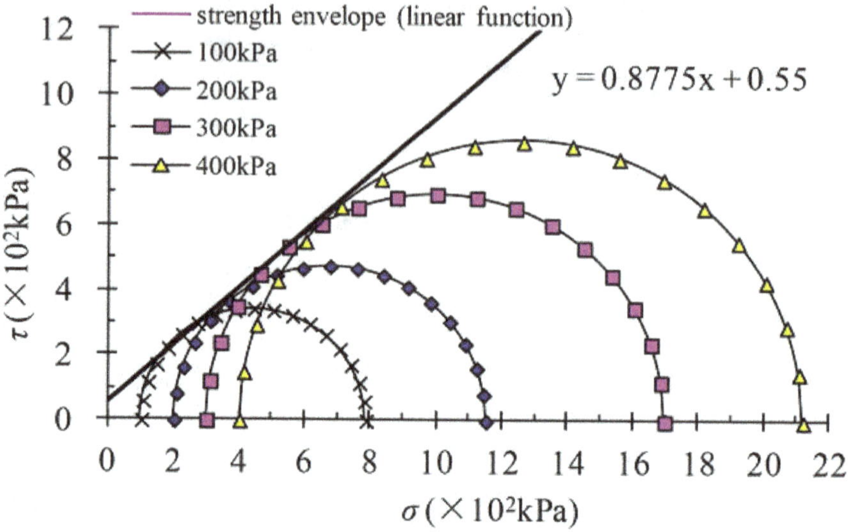

Figure 17. Mohr strength envelope of red-brown samples.

$$\tau_f = \sigma \tan 41.3° + 0.55 \text{ MPa}, \quad (5)$$

where τ f is the peak shear stress and is the normal stress. Eq. (5) shows that the cohesive strength of the samples is 55 kPa. However, because the weathered granite soil belongs to coarse grained soil, the cohesive strengths of the samples should be close to zero. It has been proved that the particles of coarse grained soil might be broken and its peak strength is nonlinear [19]. Figure 18 shows that the ratio (τ f/) is not constant but decreases when increases. In addition, the Mohr envelope is bending downward. The radius of bend reduces as the confining pressure increases, which reflects that the particle breakage is controlled by the confining pressures noticeably.

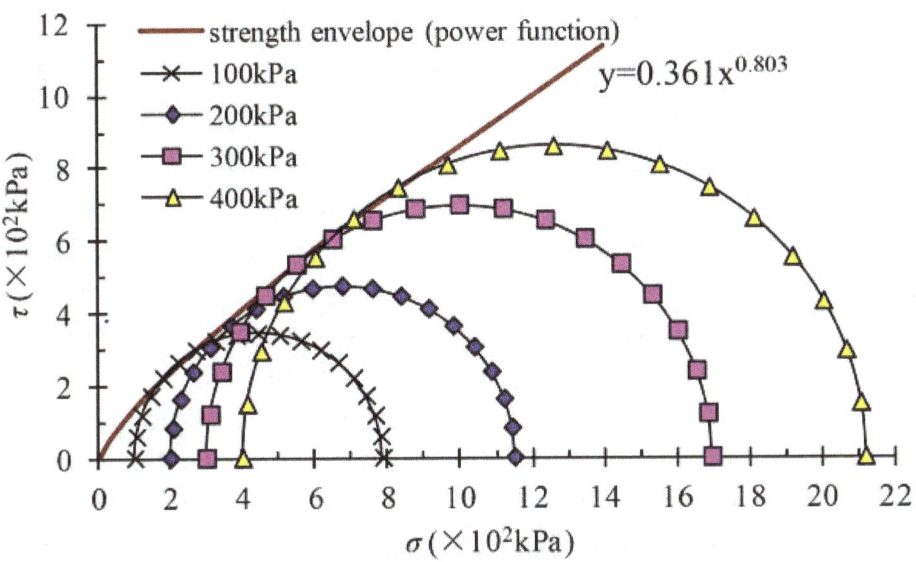

Figure 18. Actual strength envelope of red-brown samples.

According to De-Mello [20], the power function of weathered granite soil in this study can be expressed as follows:

$$\tau_f = 0.361\sigma^{0.803} \tag{6}$$

As shown in Figure 18, the fitting envelope has a good correlation, which further explains that weathered granite soil has the property of nonlinear strength because of particle breakage. The nonlinear strength characteristics of compacted weathered granite soils could be affected by rock strength, intergranular friction, occlusal effect, etc. The mechanism of the nonlinear strength could be interpreted by means of the variation of failure surface angle in previous literature [21]. Under lower confining pressure, the internal friction angle was relatively large, the bending of the strength envelope was obvious, and the samples had the obvious characteristics of shear dilation and particle breakage. With the increase of confining pressure, the internal friction angle decreased and the bending degree of strength envelope decreased gradually. It was showed that under higher confining pressure, the strength characteristics of compacted weathered granite soils were mainly dominated by particle breakage.

5.4 Shearing-dilatancy characteristics

Weathered granite soil, as coarse grained soil, has obvious shear dilation property, which was found through the CD tests in this chapter. Figure 19 shows the results of triaxial tests under difference confining pressures with respect to the relation between the stress ratio of q/p and the strain increment ratio of d v/d 1. The tests were performed according to the stress and strain increment parameters used in widely accepted models, such as Cam-clay model, where q = 1 - 3 and p = (1 + 2 3)/3. Figure 19 shows the following observations: (1) under difference confining pressures, the strain increment ratio is always larger than zero at the initial point of shear. The ratio decreases with the increase of stress ratio q/p, which indicates that the volumetric strain of weathered granite soil changes toward the volume contraction. At the same time, the contraction ratio in the initial stage of triaxial shear is larger than that in the following stage. The reason for this phenomenon is that once the soil reaches to a certain compaction rate, it gets difficult to compact further. (2) When the confining pressures change from 100 to 300 kPa, the strain increment ratio decreases as shear stress increases. The samples reached the stage transition point (d v = 0) when the strain increment ratio was equal to zero at certain level of stress. After the stage transition point, with the increase of shear force, the d v/d 1 ratio becomes negative. At this time, the volumetric strain of weathered granite soil changes toward volumetric contraction dilatancy. The aforementioned phenomena reflect the characteristics of dilatancy under lower confining pressures. The cause of the phenomena is due to the situation that the internal particles of weathered granite soil might be overturned and dislocated. (3) When the confining pressure is equal to 400 kPa, the d v/d 1 ratio is always larger than zero during shearing processes. It indicates that interparticle contact stresses in the weathered granite soil increase gradually and the particle breakage becomes obvious under larger confining pressures. At this stage, the samples were made further denser as the broken particles filled into the voids of soil and the shrinkage behavior of soil samples is obvious.

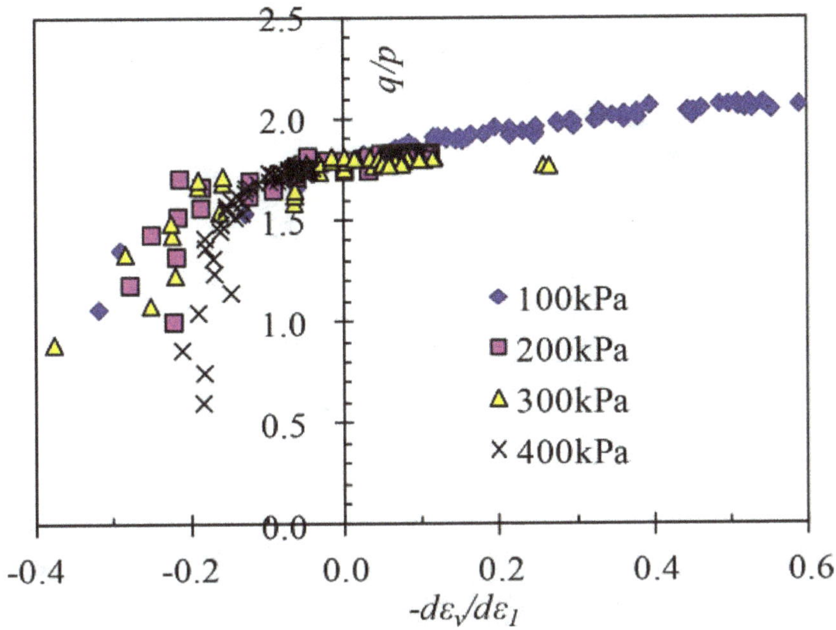

Figure 19. q/p: d v/d 1 of sample data with different confining pressures.

Based on the above test results, there must exist a confining pressure between 300 and 400 kPa that makes the volumetric strain increment to be equal to zero. In other words, between the aforementioned observations (2) and (3), there must exist a relatively balanced state when the dilatancy caused by the overturn and dislocation of soil particles is equal to the shrinkage caused by the breakage and compaction of soil particles.

The curves of 1/ 3 versus v in Figure 20 show the relationships between 1/ 3 and v. The relationships include two cases. One case is that the curve bends to left. The other case is that the curve bends to right. The volumetric strain of samples changes to the volume contraction initially and dilatancy later with the increase of the stress ratio of q/p under lower confining pressures, as shown in the former case when the curve bends to left. The volumetric strain of samples changes always toward the volume dilatancy with the increase of the stress ratio of q/p under larger confining pressures in the latter case when the curve bends to right.

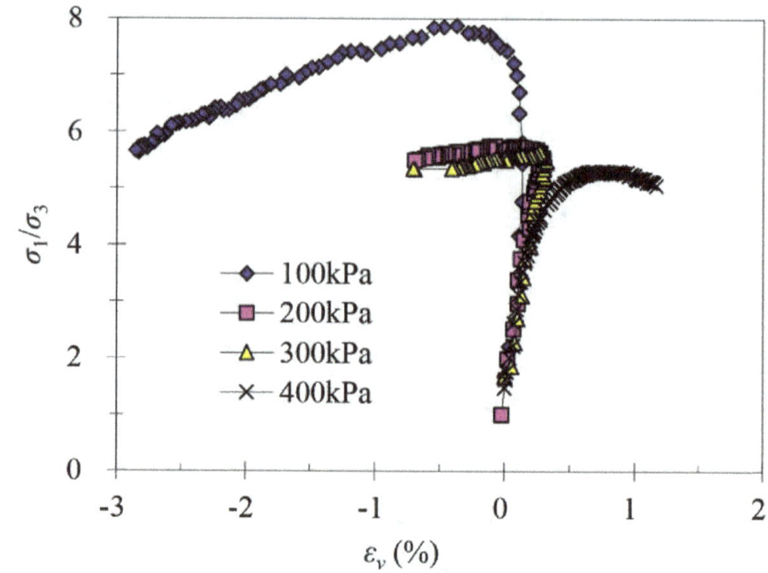

Figure 20. Curves of 1/ 3 versus v.

Figure 21 shows the relationship between peak stress ratio and peak strain increment ratio. As shown in Figure 21, an approximate linear relationship can be found between q f/p and -(d v/d 1)f. When the elastic deformation is neglected (i.e., $d\varepsilon_v/d\varepsilon_s = d\varepsilon_v^p/d\varepsilon_s^p$), M (q f/p) is equal to 1.83 based on the stress-dilatancy equation in a Cam-clay model [22].

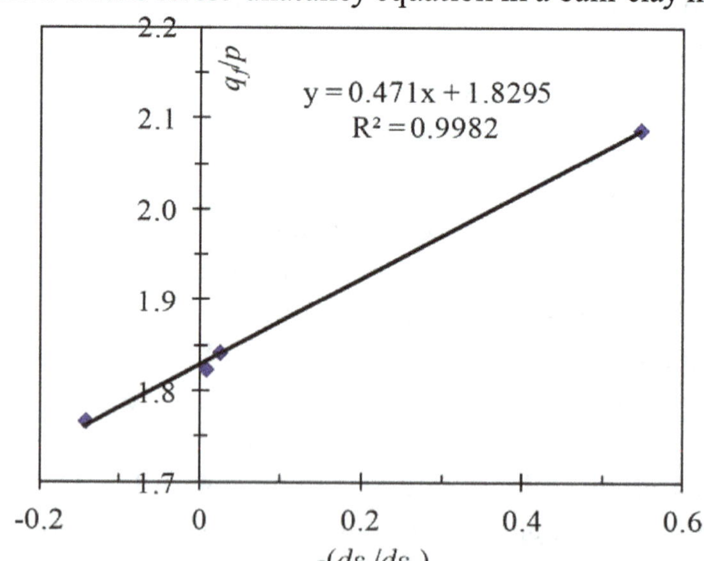

Figure 21. Curve of qf/p versus -(d v/d 1)f.

6. CONSTITUTIVE MODEL

6.1 Strength condition

Particle breakage occurs in general sand under higher confining pressure and in weathered granite soil under lower confining pressure [23]. According to the generalized nonlinear strength theory proposed by Yao et al. [24], M f and M c are expressed as follows [25, 26]:

$$M_f = M\left(\frac{p}{p_c}\right)^{-n}, \tag{7}$$

$$M_c = M\left(\frac{p}{p_c}\right)^{n}, \tag{8}$$

where M is the stress ratio at critical state, M f is the stress ratio at shear failure, M c is the stress ratio at characteristic state point, p c is the reference breaking stress, and n is the material parameter. Substituting M f = q f/p and M c = q c/p into Eqs. (7) and (8), respectively, gets the function of q f and q c as Eqs. (9) and (10) on the p-q plane, which are the exponential functions:

$$q_f = Mp_c^n p^{1-n}, \tag{9}$$

$$q_c = Mp_c^{-n} p^{1+n}. \tag{10}$$

It can be seen that the product of M f and M c is constant, as shown in Eq. (11).

$$M_f M_c = M^2, \tag{11}$$

The stress ratio at shear failure (M f) and the stress ratio at characteristic state point (M c) of granite weathered soil were calculated by a large-scale triaxial test under different confining pressures. Figure 22 shows the shape of M f and M c in p-q plane. In this figure, M is the strength line determined by Mohr-Coulomb strength criterion and M f is the actual peak strength line. From the figure, it is found that the Mohr-Coulomb strength of compacted weathered granite soil under lower confining pressure is smaller than that of the actual strength, while the strength of compacted weathered granite soil under higher confining pressure is higher than that of the actual strength. In other words, By using Mohr-Coulomb criterion, the strength of compacted weathered granite soil under lower confining pressure may be underestimated, while that of compacted weathered granite soil under higher confining pressure may be overestimated. The phenomenon of particle breakage has been found when the triaxial test for weathered granite soil in Ube, Japan, under 10–300 kPa confining pressure has been carried out by Miura et al. [23]. It can be considered that the inapplicability of the Mohr-Coulomb strength criterion for the weathered granite soil is caused by the particle breakage.

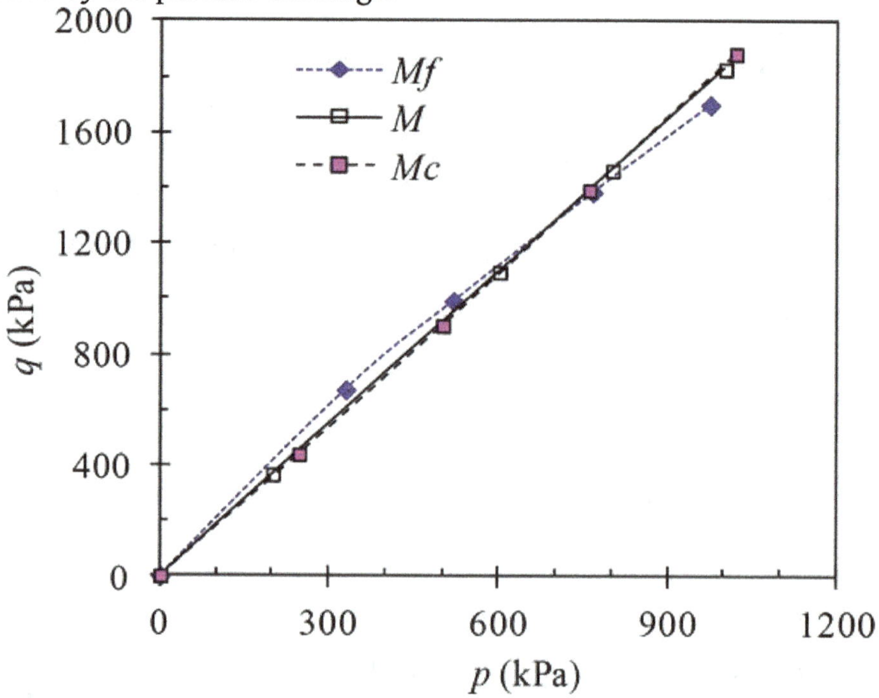

Figure 22.Shape of M f and M c on p-q plane for samples.

By fitting the stress ratio (M, M f, and M c) of weathered granite soil as shown in Figure 9, it is found that the expression by using Eqs. (12) and (13) is more accurate than by using Eqs. (7) and (8)

$$M_{\mathrm{f}} = M\left(\frac{p}{p_{\mathrm{c}}}\right)^{-4n}, \tag{12}$$

$$M_{\mathrm{c}} = M\left(\frac{p}{p_{\mathrm{c}}}\right)^{n}. \tag{13}$$

According to Eq. (14),

$$M_{\mathrm{f}}M_{\mathrm{c}} = M^2\left(\frac{p}{p_{\mathrm{c}}}\right)^{-3n}, \tag{14}$$

we can see that the product of M f and M c is not a constant, but a power function of p.

6.2 Constitutive model

6.2.1 Yield function

Based on the results of the isotropic consolidation test for Toyoura sand, Nakai [27] has established the relationship between the volumetric strain (including elastic volumetric strain and plastic volumetric strain) and the mean principal stress of sand,

$$\varepsilon_{\mathrm{v}}^{\mathrm{e}} = C_{\mathrm{e}}\left[\left(\frac{p}{p_{\mathrm{a}}}\right)^{m} - \left(\frac{p_0}{p_{\mathrm{a}}}\right)^{m}\right], \tag{15}$$

$$\varepsilon_{\mathrm{v}}^{\mathrm{p}} = (C_{\mathrm{t}} - C_{\mathrm{e}})\left[\left(\frac{p}{p_{\mathrm{a}}}\right)^{m} - \left(\frac{p_0}{p_{\mathrm{a}}}\right)^{m}\right], \tag{16}$$

where p 0 is the initial mean principal stress, p a (=101.3 kPa) is the atmospheric pressure, m is a coefficient for sand, C t is the compression index, and C c is the swelling index. Based on the Eqs. (15) and (16), the yield function of weathered granite soil is constructed in this chapter.

When considering the particle breakage of soil, the stress-dilatancy equation in the modified Cambridge model was changed by Yao et al. [28] to the following form:

$$\frac{d\varepsilon_{\mathrm{v}}^{\mathrm{p}}}{d\varepsilon_{\mathrm{s}}^{\mathrm{p}}} = \frac{M_{\mathrm{c}}^2 - \eta^2}{2\eta}, \tag{17}$$

where psεsp is the plastic shear strain and ηη is q/p.

The orthogonality condition is

$$-\frac{d\varepsilon_v^p}{d\varepsilon_s^p}=\frac{dq}{dp}, \tag{18}$$

The expression of the plastic potential function, which is obtained through the differential equation, composed of Eqs. (17) and (18), is expressed as

$$f=(2n+1)\frac{p_c^{2n}}{M^2}\frac{q^2}{p}+p^{2n+1}-p_x^{2n+1}=0. \tag{19}$$

According to Nakai's research [27], p x is expressed as

$$p_x=\left(\frac{p_a^m}{C_t-C_e}\varepsilon_v^p+p_0^m\right)^{\frac{1}{m}}. \tag{20}$$

Substituting the Eq. (20) into the Eq. (19), we get

$$f=\frac{C_t-C_e}{p_a^m}\left\{\left[\frac{(2n+1)p_c^{2n}}{M^2}\frac{q^2}{p}+p^{2n+1}\right]^{\frac{m}{2n+1}}-p_0^m\right\}-\varepsilon_v^p=0. \tag{21}$$

The above expression is the yield function of weathered granite soil. In the function, pvεvp is an independent variable and can be used as the hardening parameter of the constitutive model of weathered granite soil.

6.2.2 Hardening parameter

On the basis of existing research results, Yao et al. [29] proposed the hardening parameters considering particle breakage as follows:

$$H=\int\frac{M_c^4}{M_f^4}\frac{(M_f^4-\eta^4)}{(M_c^4-\eta^4)}d\varepsilon_v^p. \tag{22}$$

Simplifying the differential form of Eq. (21), we obtain

$$d\varepsilon_v^p=\frac{M_f^4}{M_c^4}\frac{(M_c^4-\eta^4)}{(M_f^4-\eta^4)}dH. \tag{23}$$

Eq. (23) is used as a hardening parameter to construct the constitutive model of weathered granite soil.

6.2.3 Constitutive relation

By using the theory of plastic displacement potential and the associated flow rule and using the basic assumptions in the Cambridge model, after d pvdεvp in the yield function, Eq. (21), was replaced by the hardening parameter H considering particle breakage in Eq. (22), we obtain,

$$\frac{\partial f}{\partial p} = \frac{C_t - C_e}{p_a^m} m \left[\frac{(2n+1)p_c^{2n}}{M^2} \frac{q^2}{p} + p^{2n+1} \right]^{\frac{m-2n-1}{2n+1}} \left(p^{2n} - \frac{p_c^{2n}}{M^2} \frac{q^2}{p^2} \right), \tag{24}$$

$$\frac{\partial f}{\partial q} = \frac{C_t - C_e}{p_a^m} m \left[\frac{(2n+1)p_c^{2n}}{M^2} \frac{q^2}{p} + p^{2n+1} \right]^{\frac{m-2n-1}{2n+1}} \left(\frac{2p_c^{2n}}{M^2} \frac{q}{p} \right), \tag{25}$$

$$\frac{\partial f}{\partial H} = -1 \tag{26}$$

According to the definition of generalized deviant stress, the plastic volumetric strain can be expressed as follows by means of a series of transformations of Eqs. (24)–(26):

$$\begin{aligned} d\varepsilon_v^p = {} & \frac{m(C_t - C_e)}{p_a^m} \frac{M_f^4 \left(M_c^4 - \eta^4\right)}{M_c^4 \left(M_f^4 - \eta^4\right)} \left(dp + \frac{2\eta}{M_c^2 - \eta^2} dq \right) \\ & \left[\frac{(2n+1)p^{2n+1}}{M_c^2} \eta^2 + p^{2n+1} \right]^{\frac{m-2n-1}{2n+1}} \left(p^{2n} - \frac{p^{2n}}{M_c^2} \eta^2 \right). \end{aligned} \tag{27}$$

The stress-dilatancy equation is expressed in the following form from Eqs. (17) and (27)

$$\begin{aligned} d\varepsilon_s^p = {} & \frac{m(C_t - C_e)}{p_a^m} \frac{M_f^4 \left(M_c^2 + \eta^2\right) 2\eta}{M_c^4 \left(M_f^4 - \eta^4\right)} \left(dp + \frac{2\eta}{M_c^2 - \eta^2} dq \right) \cdot \\ & \left[\frac{(2n+1)p^{2n+1}}{M_c^2} \eta^2 + p^{2n+1} \right]^{\frac{m-2n-1}{2n+1}} \left(p^{2n} - \frac{p^{2n}}{M_c^2} \eta^2 \right) \end{aligned} \tag{28}$$

Equations (26) and (27) are the plastic strain expression of weathered granite soil considering particle breakage.

In three-dimensional axisymmetry, the following expressions can be obtained from the generalized Hook's law:

$$d\varepsilon_v^e = \frac{3(1-2v)}{E} dp \tag{29}$$

$$d\varepsilon_s^e = \frac{2(1+v)}{3E} dp \tag{30}$$

where E is the elastic modulus and v is the Poisson's ratio.

After taking the derivative of p in Eq. (15), the elastic volumetric strain can be expressed as

$$\mathrm{d}\varepsilon_{\mathrm{v}}^{\mathrm{e}} = \frac{mC_{\mathrm{e}}p^{m-1}}{p_{\mathrm{a}}^{m}}\mathrm{d}p \tag{31}$$

Combining Eqs. (29)–(31), the elastic shear strain is obtained in the following form,

$$\mathrm{d}\varepsilon_{\mathrm{s}}^{\mathrm{e}} = \frac{2(1+v)mC_{\mathrm{e}}p^{m-1}}{9(1-2v)p_{\mathrm{a}}^{m}}\mathrm{d}p \tag{32}$$

According to elastoplastic mechanics, the total volumetric strain and the total shear strain are, respectively, expressed as

$$\begin{gathered}\mathrm{d}\varepsilon_{\mathrm{v}} = \mathrm{d}\varepsilon_{\mathrm{v}}^{\mathrm{e}} + \mathrm{d}\varepsilon_{\mathrm{v}}^{\mathrm{p}} = \frac{m(C_{\mathrm{t}}-C_{\mathrm{e}})}{p_{\mathrm{a}}^{m}}\frac{M_{\mathrm{f}}^{4}}{M_{\mathrm{c}}^{4}}\frac{(M_{\mathrm{c}}^{4}-\eta^{4})}{(M_{\mathrm{f}}^{4}-\eta^{4})}\left(\mathrm{d}p+\frac{2\eta}{M_{\mathrm{c}}^{2}-\eta^{2}}\mathrm{d}q\right)\cdot \\ \left[\frac{(2n+1)p^{2n+1}}{M_{\mathrm{c}}^{2}}\eta^{2}+p^{2n+1}\right]^{\frac{m-2n-1}{2n+1}}\left(p^{2n}-\frac{p^{2n}}{M_{\mathrm{c}}^{2}}\eta^{2}\right)+\frac{mC_{\mathrm{e}}p^{m-1}}{p_{\mathrm{a}}^{m}}\mathrm{d}p\end{gathered} \tag{33}$$

$$\begin{gathered}\mathrm{d}\varepsilon_{\mathrm{s}} = \mathrm{d}\varepsilon_{\mathrm{s}}^{\mathrm{e}} + \mathrm{d}\varepsilon_{\mathrm{s}}^{\mathrm{p}} = \frac{m(C_{\mathrm{t}}-C_{\mathrm{e}})}{p_{\mathrm{a}}^{m}}\frac{M_{\mathrm{f}}^{4}}{M_{\mathrm{c}}^{4}}\frac{(M_{\mathrm{c}}^{2}+\eta^{2})2\eta}{(M_{\mathrm{f}}^{4}-\eta^{4})}\left(\mathrm{d}p+\frac{2\eta}{M_{\mathrm{c}}^{2}-\eta^{2}}\mathrm{d}q\right)\cdot \\ \left[\frac{(2n+1)p^{2n+1}}{M_{\mathrm{c}}^{2}}\eta^{2}+p^{2n+1}\right]^{\frac{m-2n-1}{2n+1}}\left(p^{2n}-\frac{p^{2n}}{M_{\mathrm{c}}^{2}}\eta^{2}\right)+\frac{2(1+v)mC_{\mathrm{e}}p^{m-1}}{9(1-2v)p_{\mathrm{a}}^{m}}\mathrm{d}q.\end{gathered} \tag{34}$$

The Eqs. (33) and (34) are the constitutive relation of the weathered granite soils considering particle brakeage.

The seven parameters in the constitutive model of weathered granite soil are C e, C t, m, M, p c, n, and . Except for Poisson's ratio , the other six parameters can be obtained by conventional triaxial test. The value of is assumed to be 0.3. C e, C t, and m can be obtained by isotropic compression and unloading test, and M, p c, and n can be obtained by consolidated drained triaxial test.

7. CONCISE SUMMARY

We did regular physical and mechanics tests as well as large-scale triaxial tests in this chapter to look into the compaction, bearing, strength, shearing-dilatancy, and constitutive model of compacted worn granite soil. The main findings that this chapter's study led to are summed up below:

- With an increase in burying depth of weathered granite, the geological year's parameter (m) decreases by power function, but the geometric progression constants (r) increase by power function. The exponent of power function in this chapter can be used to evaluate the weathering process of weathered granite [11].
- The particle-breaking characteristics of weathered granite soils are obviously influenced by many factors such as particle gradation, mineral content, blows per layer, and stress level.
- The pure weathered granite soil shows similar compaction characteristics of sands. The compression and bearing characteristics of weathered granite soil vary significantly when the clay content ratio changes. The weathered granite soil in this chapter can be used as the filling material.
- The experimental results show that the increase in peak deviatoric stress q f due to an increase in mean stress p is observed as nonlinear under lower confining pressure because of the existence of particle breakage. It is found that the product of the state stress ratio and peak stress ratio is not a constant but a power function of an average main stress. The research results are helpful to understand the law of the long-term degradation of subgrade performance due to particle breakage of weathered granite soil.

Since wet conditions can change how embankments change after they are built, more research should be done on dynamic methods for measuring particle breakage and long-term behavior degradation analysis of subgrade due to more weathering and particle breakage [7, 11, 30, 31]. Because it has unique mechanical qualities, it is hard to fully understand how well weathered granite soil works as a road base. More research is needed to figure out how different types of weathered granite soils (like coarse-grained weathered granite soil and fine-grained weathered granite soil) behave mechanically when they are compacted. This includes looking at how particles break apart and the mechanical properties of each type of soil. It is still hard to figure out how to measure particle breakage and how it changes during triaxial compression tests [32]. Because of this, more research needs to be done on dynamic measurement methods, the mathematical description of particle breakage, and how particle breakage affects long-term behavior degradation [33].

AT A GLANCE

There are a lot of weathered granite soils in mountainous places all over the world. Over time, civil engineering will get better, which will allow for the building of more roads, railways, dams, and other infrastructure in mountainous places. Mountainous places may not have enough building materials, but they can make better use of weathered granite soils, which are a special kind of geomaterial. Weathered granite soil has its own unique physical and mechanical qualities, such as breaking down easily and breaking up into small pieces. There were a lot of field studies and lab tests (such as X-ray diffraction, sieving, heavy compaction, and large-scale triaxial) that went into this part. We've talked about the weathering process, the factors that affect particle breakage, and the mechanical features of compacted granite that has been weathered. The findings show that particle size, mineral content, number of hits per layer, and stress level all have a big impact on how particles break down in weathered granite soils. It was found that the product of the stress ratio at shear failure M f and the stress ratio at characteristic state point M c is not a constant. Instead, it is a power function of the average main stress p because particles break. This is how the fundamental model of weathered granite soils was put forward.

REFERENCES

1. Lu F, Sang L. Petrology. Beijing: Geological Publishing House; 2002. p. 120

2. Wu N, Zhao C, Hou W. Research on the cause of formation, distribution and engineering characteristics of the granite residual soil. Journal of Pingdingshan Institute of Technology. 2004;13:1-4

3. Writing Committee of Handbook of Engineering Geology (WCHEG). Handbook of Engineering Geology. 4th ed. Beijing: China Building Industries Press; 2007. p. 125

4. Fukumoto T. A grading equation for decomposed granite soil. Soils and Foundations. 1990;30:27-34. DOI: 10.3208/sandf1972.30.27

5. Yu F. Particle breakage and the drained shear behavior of sands. International Journal of Geomechanics. 2017;17:04017041. DOI: 10.1061/(ASCE)GM.1943–5622.0000919

6. Niu X, Yao Y, Chen Z. The strength and constitutive model of compacted weathered granite soils in Lüliang mountains. Rock and Soil Mechanics. 2017;38:2833-2840. DOI: 10.16285/j.rsm.2017.10.008

7. Konrad JM, Salami Y. Particle breakage in granular materials—A conceptual framework. Canadian Geotechnical Journal. 2018;55:710-719. DOI: 10.1139/cgj-2017-0224

8. Ministry of Transportation of the People's Republic of China (MOTPRC)Test Methods of Soil for Highway Engineering (JTG E40-2007). Beijing: People's Communications Press; 2007

9. China Electronics Corporation (CEC). Code for Coarse-Grained Soil Tests for Hydropower and Water Conservancy Engineering (DL/T 5356-2006). Beijing: China Power Press; 2006

10. Geng Y, Yang C, Song B, Wan Y. Post-orogenic granites with an age of 1800 ma in Luliang area, North China Craton: Constraints from isotopic geochronology and geochemistry. Geological Journal of China Universities. 2004;10(4):477-487. DOI: 10.3969/j.issn.1006-7493.2004.04.001

11. Niu X, Yao Y, Sun Y, Luo Z. Weathering process of in situ granite and particle breakage characteristics of compacted weathered granite. Applied Sciences. 2018;87:1108. DOI: 10.3390/app8071108

12. Hardin BO. Crushing of soil particles. Journal of Geotechnical Engineering. 1985;111:1177-1192. DOI: 10.1061/(ASCE)0733-9410(1985)111:10(1177)

13. Leslie DD. Large scale triaxial tests on gravelly soils. In: Proceedings of the 2nd Pan-American Conference on Soil Mechanics and Foundation Engineering; 16-24 July 1963; Rio de Janeiro, Brazil. Vol. 1. São Paulo, Brazil: Associacao Brasileira de Mecanica dos Solos; 1963. pp. 181-202

14. Leslie DD. Shear Strength of Rockfill, Physical Properties. Sausalito, CA, USA: Engineering Study No. 526; South Pacific Division, Corps of Engineers Laboratory; 1975. p. 124

15. Marsal RJ. Discussion of shear strength. In: Proceedings of the 6th International Conference on Soil Mechanics and Foundation Engineering; Montreal, QC, Canada. Vol. 3. 1965. pp. 310-316

16. Lee KL, Farhoomand I. Compressibility and crushing of granular soils in anisotropic triaxial compression. Canadian Geotechnical Journal. 1967;4:68-86. DOI: 10.1139/t67-012

17. Niu X, Sun Y, Yao Y, Han J. Basic properties and weathering process of strong weathered granite. Japanese Geotechnical Society Special Publication. 2015;1:8-33. DOI: 10.3208/jgssp.CPN-07

18. Lee IK, Coop MR. The intrinsic behaviour of a decomposed granite soil. Geotechnique. 1995;45:117-130. DOI: 10.1680/geot.1995.45.1.117

19. Liu M, Gao Y, Liu H, Chen Y. Large-scale triaxial test study on deformation and strength characteristics of rockfill materials. Chinese Journal of Rock Mechanics and Engineering. 2003;22:1104-1111. DOI: 10.3321/j.issn:1000-6915.2003.07.011

20. De-Mello VFB. Reflections on decisions of practical significance to embankment dam construction. Géotechnique. 1977;27:281-355. DOI: 10.1680/geot.1977.27.3.281

21. Vesic AS, Clough GW. Behavior of granular materials under high stresses. Journal of the Soil Mechanics and Foundations Division. 1968;94:661-688

22. Roscoe KH, Burland JB. On the generalized stress-strain behavior of “wet” clay. In: Engineering Plasticity. Cambridge, U.K.: Cambridge Univ. Press; 1968. pp. 535-609

23. Miura N, O-Hara S. Particle-crushing of a decomposed granite soil under shear stresses. Soils and Foundations. 1979;19:1-14. DOI: 10.3208/sandf1972.19.3_1

24. Yao Y, Lu D, Zhou A. Generalized non-liner strength theory and its transformed stress space. Science in China, Series E Technological Sciences. 2004;47:691-709. DOI: 10.1360/04ye0199

25. Yao Y, Zhou A, Lu D. Extended transformed stress space for geomaterials and its application. Journal of Engineering Mechanics. 2007;133:1115-1123. DOI: 10.1061/(asce)0733-9399(2007)133:10(1115)

26. Yao Y, Sun D, Matsuoka H. A unified constitutive model for both clay and sand with hardening parameter independent on stress path. Computers and Geotechnics. 2008;35:210-222. DOI: 10.1016/j.compgeo.2007.04.003

27. Nakai T. An isotropic hardening elastoplastic model for sand considering. Soils and Foundations. 1989;29:119-137

28. Yao Y, Matsuoka H, Sun D. A unified elastoplastic model for clay and sand with the SMP criterion. In: Proceedings of the 8th Australia-New Zealand Conference on Geomechanics; Hobart, Australia. Hobart: Australian Geomechanics Society; 1999. pp. 997-1003

29. Yao Y, Sun D, Luo T, Matsuoka H. A simple 3-D constitutive model for both clay and sand. Chinese Journal of Geotechnical Engineering. 2002;24:240-246. DOI: 10.3321/j.issn:1000-4548.2002.02.026

30. Wang W, Coop MR. An investigation of breakage behaviour of single sand particles using a high-speed microscope camera. Géotechnique. 2016;66:1-15. DOI: 10.1680/jgeot.15.P.247

31. Niu X, Xie H, Sun Y, Yao Y. Basic physical properties and mechanical behavior of compacted weathered granite soils. International Journal of Geomechanics. 2017;17:04017082. DOI: 10.1061/(ASCE)GM.1943-5622.0000983

32. Indraratna B, Thakur PK, Vinod JS, Salim W. Semiempirical cyclic densification model for ballast incorporating particle breakage. International Journal of Geomechanics. 2012;12:260-271. DOI: 10.1061/(ASCE)GM.1943-5622 .0000135

33. Chazallon C, Hornych P, Mouhoubi S. Elastoplastic model for the long-term behavior modeling of unbound granular materials in flexible pavements. International Journal of Geomechanics. 2006;6:279289. DOI: 10.1061/(asce)1532-3641(2006)6:4(279)

GLOSSARY

AASHTO Soil Classification: A soil classification system usually used for highway design and construction.

Active earth pressure: The horizontal stress exerted by a mass of soil on a retaining wall as the wall moves away from the soil. **Active (Rankine) zone:** The area behind a retaining structure that is above the failure plane.

Activity: The ratio of plasticity index to percent by weight of clay. This property is correlated with the type of clay.

Adhesion: The shear resistance between soil and a structure (e.g. steel, concrete or timber piles; along a retaining wall)

Air-voids ratio; The ratio of the volume of air to the total volume of a mass of soil.

Allowable bearing capacity: The bearing pressure that can be allowed on a foundation soil, usually to limit settlements.

Alluvial soils: Soils deposited in a valley or slightly graded area by transporting sediments through a mountain river or streams.

American Association of State Highway Transportation Officials' Classification System (AASHTO): A system of soil classification based on grain size, liquid limit and plasticity of soils.

Angle of internal friction: For a given soil, the angle on the graph of the shear stress and normal effective stresses at which shear failure occurs.

Angle of repose: The maximum angle, just before failure, of a slope composed of granular material.

Angle of shearing resistance: The ratio of effective shear and normal stresses mobilized at any state prior to failure.

Angle of slip plane: The angle referred to horizontal of a plane or other surface along which a discontinuous slip or rupture may occur.

Angle of wall friction: The angle of friction between soil and the surface of a retaining wall or bottom side of a foundation.

Angular distortion: The ratio between the relative deflection between two points in a foundation and the distance between them.

Anisotropic: A mass of soil having different properties in different directions. Primarily refers to permeability or stress-strain characteristics.

Aquifer: A stratum of soil with relatively high permeability; a water-bearing stratum of rock or soil.

Artesian: A condition that exists when the water table piezometric surface lies above the ground level.

At-rest earth pressure: The horizontal stress developed in a mass of soil loaded in conditions of zero horizontal strain. The retaining wall neither moves away or towards the soil.

Atterberg Limits: The water contents of a soil mass corresponding to the transition between a solid, semi-solid, plastic solid or liquid. Laboratory test used to distinguish the plasticity of clay and silt particles.

Axial strain: Direct strain meaured along an axis of a triaxial test sample.

Axial stress: Total or effective stress (Both confining and vertical stresses combined) acting along an axis of a triaxial test sample.

Bearing capacity: The ability of the underlying soil to support the foundation loads without shear failure.

Bearing capacity factors: Empirically derived factors used in a bearing capacity equation that usually correlates with the angle of internal friction of the soil.

Bearing pressure: The total stress transferred from the structure to the foundation, then to the soil below the foundation.

Bedrock: Strong rock underlying surface deposits of soil and weathered rock.

Bentonite: Sodium montmorillonite.

Borrow: A site that is used to supply soils for earthwork construction (i.e. gravel pit).

Boulders: Soil particles over 12 inches (305 mm) in size.

Boussinesq equation: An equation used to determine the increase in vertical pressure at a particular depth that is caused by an application of a point load at a given surface.

Braced excavation: The use of bracing to laterally support the side-walls of temporary trenches or cuts.

British Soil Classification System: A system of soil classification based on size, consistency and structure.

Bulk density: Soil density. The total mass of water and soil particles contained in a unit volume of soil. (pcf).

Bulk unit weight: The total weight of water and soil particles contained in a unit volume of soil.

Buoyant Density: Also, submerged density. Difference between the total density and the density of water. Buoyant density = Soil density - Density of water.

Caisson: A component of a particular foundation system.

California Bearing Ratio (CBR): A laboratory test that is used to determine the suitability of of a soil for use as a subbase in a pavement section.

Capillary rise: The height to which water will rise above the water table due to negative pore water pressure or capillary action of the soil.

Capillary stresses: Pore water pressures less than atmospheric values produced by surface tension of pore water acting on the meniscus formed in void spaces between soil particles.

Circular Footing: Isolated/ spread footing that is circular shaped. Usually a shallow footing.

Clay: Soil particles which are finer (smaller) than 0.002 mm in size.

Coarse-grained soils: Soils with more than 50% by weight of grains retained on the #200 sieve (0.075mm).

Cobbles: Soil particles between 3 inches (76 mm) and 12 inches (305 mm) in size.

Coefficient of active earth pressure: The ratio of the minimum horizontal effective stress of a soil to the vertical effective stress at a single point in a soil mass retained by a retaining wall as the wall moves away from the soil.

Coefficient of compressibility: The ratio of void ratio difference to the effective pressure difference of two different loadings during primary consolidation.

Coefficient of consolidation: The rate of change of volume during primary consolidation. Units: ft^2/day, m^2/s.

Coefficient of curvature: Also curvature coefficient. A measure of the shape parameter obtained from a grain size distribution curve. $C_D=(D_{30})^2/D_{10}D_{60}$.

Coefficient of earth pressure at rest: The ratio of horizontal effective stress of a soil to the vertical effective stress at a specific point in a soil mass behind a retaining wall when the wall does not move (loaded in conditions of zero horizontal strain).

Coefficient of friction: The ratio between the tangential force required to cause a body, such as a foundation surface or a retaining wall, to slide along a plane and the normal force between the body and the plane.

Coefficient of passive earth pressure: The ratio of the maximum horizontal effective stress of a soil to the vertical effective stress at a specific point in a soil mass behind a retaining wall as the wall moves toward the soil. See the link to Retaining Walls for equations and calculations.

Coefficient of secondary consolidation: The change in volumetric strain per a logarithmic cycle of time after primary consolidation is complete.

Coefficient of uniformity: Also uniformity coefficient. A measure of the slope of a grain size distribution curve, and therefore the uniformity of the soil. $C_u = D_{60}/D_{10}$.

Cofferdam: A temporary structure used to enclose a construction area, and prevent soil or water from entering the construction area.

Cohesionless soils: Granular soils (sand and gravel type) with values of cohesion close to zero.

Cohesive soils: Clay type soils with angles of internal friction close to zero. Cohesion is the force that holds together molecules or like particles within a substance.

Colluvial soils: Soils deposited at the base of foot-hills via gravity or erosion.

Compaction: Volume change in soils which air is expelled from the voids, but with the water content remaining constant. Compaction may occur due to vibration, and self-weight. In construction, compaction is achieved by rolling, tamping or vibrating fill soils.

Compression index: (Also, compressibility index) The logarithmic slope of the primary consolidation curve. The slope of the normal compression line and critical state line of the Casagrande Method of consolidation curves.

Compression Test: The compression test is performed in the laboratory using a relatively undisturbed ring sample. The sample is loaded with expected building pressures to estimate the amount of compression the soil undergoes. The sample is then inundated with water in order to measure additional compression or swelling.

Cone resistance: The resistance force divided by the end area of the cone tip, measured during the cone penetration test.

Cone penetration test: A penetration test in which a cone that has a 60º point is pushed into the ground at a continuous rate. Resistance is measured by correlating the depth penetrated with the force applied.

Confined aquifer: An aquifer that is contained between two stratifications of low permeability soil or rock.

Consistency: Water content of clay relative to the Atterberg limits.

Consistency index: A measure of the relationship between the current water content and the consistency limits.

Consolidation: Volume change due to dissipation of excess pore pressure from static loads.

Consolidation (settlement): The settlement of a foundation due to squeezing out of water from the pores as the soil comes to equilibrium with the applied loads.

Consolidation test: A laboratory test in which results are used to predict consolidation of a soil under applied structural loads. Also known as confined compression tests and oedometer tests.

Continuous footing: Also see "Strip footing." A horizontally long footing supporting a wall. Usually a shallow footing.

Coulomb earth pressure theory: An earth pressure theory that includes friction between the soil and retaining structure, and assumes that failure occurs along a flat plane behind the retaining structure at an angle that is in part derived from the angle of internal friction.

Coulomb's equation: Named after Charles Augustin Coulomb, (1736-1806) An equation relating the shear strength of soil to the normal effective stress on the failure plane.

Counterfort walls: Type of retaining wall that depends on tension ribs between the stem and the heel in order to resist flexure and overturning.

Creep: Time dependent deformations which occurs in soil at constant effective stress without changes in volume and pore water pressure.

Critical circle: In a slope stability analyses, the slip circle that corresponds to the lowest factor of safety.

Critical ground slope angle: The angle of the ground slope that corresponds to a factor of safety of 1.0 relative to the slope stability.

critical height: The height of a slope that corresponds to a factor of safety of 1.0 relative to slope failure.

Critical hydraulic gradient: The hydraulic gradient at which effective stresses becomes zero. Upward seepage.

Cyclic stress ratio: A numerical rating of the potential for liquefaction in sands.

Darcy's law: (After H.P.G. Darcy, 1856) The formula used for laminar flow of water through porous saturated soils. Velocity of flow = hydraulic conductivity X hydraulic gradient.

Degree of consolidation: The proportion of consolidation that has occurred after a given elapsed time.

Degree of saturation: The proportion of the volume of water to the total volume of voids of a given mass of soil.

Density: Theratioofthetotalmasstothetotalvolumeofaunitofsoil.Usually expressed as a unit weight where weight is interchanged with mass. Units: lbm/ft^3, kg/m^3.

Density index: Also, relative density. The density of a granular soil relative to the minimum and maximum densities achieved for that particular soil.

Density of soil grains: The average density of the mineral or rock of which the soil particles are composed.

Density of water: The density of water will vary with temperature and pressure. Values used for soils analysis are 62.4 lbm/ft^3, 9.81 kN/m^3 or 1000 kg/m^3. A value of 64.0 lbm/ft^3 is the value for sea water.

Depth factor: One of the terms in the bearing capacity equation that relates to depth of the foundation. The ratio between the depth of a slip circle below the top of a slope and the height of the slope.

Desiccation: The process of shrinkage or consolidation of the fine-grained soil produced by increase of effective stresses in the grain skeleton accompanying the development of capillary stresses in the pore water.

Deviator stress: The difference between the axial and radial stresses of a triaxial test sample.

Dewater: Removal of water from a job site. Usually by pumping from excavations.

Differential settlement: The vertical displacement due to settlement of one point in a foundation with respect to another point of the foundation.

Direct shear test: Laboratory test used to determine the relationship of shear strength to consolidation stress. Strength characteristics that are estimated from this test includes cohesion and angle of internal friction.

Direct strain: The ratio of the change in length to the original length of a soil mass.

Downdrag: Negative skin friction. Forces induced on deep foundations resulting from downward movement of adjacent soil relative to the foundation element.

Drained loading: Loading which is slow enough for the water to drain from the soil as the total stresses increase. Pore pressure will not change, and volume will with loading.

Drawdown: The magnitude of the lowering of a water table, usually near a well being pumped.

Dredging: Removing soils from a sea, river or lake bed in order to deepen the waterway for water travel.

Dry density: The ratio of the mass of the solids (soil grains) to the total unit volume of soil. Units: lbm/ft^3, kg/m^3.

Dry unit weight: The weight of solids (soil grains) to the total unit volume of soil. Units lb/ft^3, kN/m^3.

Dynamic compaction: The use of high-energy impact to densify loose granular soils.

Earth pressure: The force per unit area exerted by soil on a retaining wall.

Earth pressure coefficients: Coefficients used in determining earth pressure. Components may include angle of internal friction, friction between the soil and wall face, angle of the wall face, and angle of the sloping backfill.

Effective stress: The portion of the total stress that is supported through grain-to-grain contact of the soil. The stress in a soil mass that is effective in causing volume changes and in mobilizing the shear strength arising from friction. The difference between the total stress and the pore water pressure. Effective stress = Total stress - Pore water pressure.

Efficiency of a pile: Also, pile efficiency. For a given pile in a group of piles, the ratio of the average ultimate load in the group to the individual ultimate load on the given pile.

Elastic deformation: Deformation caused in a soil due to a change in loading, and the soil recovers completely when the load is removed.

Embankment: Usually referred to a 'built-up' section of soil (engineered fill) as for roads or dams.

Engineering properties: Engineering parameters of a soil such as permeability, shear strength and consolidation. Different from index properties.

Engineered fill: Soils used as fill, such as retaining wall backfill, foundation support, dams, slopes, etc.. that are to be placed in accordance to engineered specifications. These specifications may delineate soil grain-size, plasticity, moisture, compaction, angularity, and many other index properties depending on the application.

Eolian Soil: Soil deposits that have been transported by wind.

Equipotential: For a flow net, lines connecting points of equal total head. Equipotential lines are usually drawn so that the interval, or equipotential drop, is constant. Equipotentials intersect flow lines and impermeable boundaries at right angles.

Equivalent fluid pressure: Horizontal pressures of soil, or a combination of soil and water, which increase linearly with depth and are equivalent to those that would be produced by a heavy fluid of a selected unit weight.

Excess pore pressure: That increment of pore water pressures greater than hydro-static values, produced by consolidation stresses in compressible materials or by shear strain; excess pore pressure is dissipated during consolidation.

Exit gradient: The hydraulic gradient near an exposed surface through which seepage is moving.

Expansive Clays: Also, Reactive Clays. Clays that are sensitive to water, causing them to swell or expand.

Factor of safety: The ratio of a limiting value of a quantity to the design value of that quantity.

Failure envelope: For a given soil, the graph of the shear stress and normal effective stresses at which shear failure occurs.

Fault: A shear fracture in a rock mass along which movement has taken place.

Field density test: Also In-place density test. Field testing that determines density of compacted fill to verify that it meets specifications. Types of tests may include sand cone, rubber balloon or nuclear densiometer.

Fine-grained soils: Silt and clay soils. Soils containing particles smaller than No. 200 sieve or 0.075 mm in size according to the Unified Soil Classification System.

Fines content (fraction): Soil grains smaller than No. 200 sieve (0.075 mm), e.g. Clay and Silt.

Fissured clay: A clay having an internal network of narrow cracks or separations, in which the width and depth tends to increase upon drying.

Flow line: The path water will follow traveling from high head to low head in a seepage flow analysis.

Flow net: A graphical analysis of seepage flow in a mass of soil to estimate flow quantities and pore pressures.

Flow quantity: The total volume of water flowing in a seepage analysis.

Flow rate: The ratio of total volume of water flowing to a particular unit of time.

Flow slide: Shear failure in which a soil mass moves over a relatively long distance in a fluid-like manner, occurring rapidly on flat slopes in loose, saturated, uniform sands, or in highly sensitive clays.

Flow velocity: The velocity of water flow through a soil.

Footing: An enlargement at the base of a foundation that is designed to transmit forces to the soil.

Foundation: A component of an engineered structure that transmits the structure's forces into the soil or rock that supports it.

Founding depth: The depth below the ground surface where the base of a foundation is located.

Friction pile: A pile that derives the majority of its load bearing ability from the skin friction between the soil and the pile.

Frost jacking: A pile forced upward due to freezing ground moving upward or repeated freeze-thaw events.

Geotextiles: A synthetic fabric used to stabilize soils, retain soils, prevent the mixing of dissimilar soils, provide a filtering function, pavement support, subgrade reinforcement, drainage, erosion control and silt containment.

Grain size distribution curve: A curve drawn on a log scale to represent the distribution of particle sizes in a soil.

Gravity walls: Retaining walls which depend upon their self weight to provide stability against overturning and sliding; usually made of a high bulk structure.

Homogenous earth dam: An earth dam whose embankment is formed of one soil type without a systematic zoning of fill materials.

Homogenous soils: A mass of soil where the soil is of one characteristic having the same engineering and index properties.

Horizontal strain: Strain measured in a horizontal direction.

Horizontal stress: Total or effective stress acting in a horizontal direction.

Hveem's Resistance Value Test: Laboratory test that evaluates the resistance of a re-molded soil sample to be used in a pavement section.

Hydraulic conductivity: Also, coefficient of permeability. The constant average discharge velocity of water passing through soil when the hydraulic gradient is equal to 1.0. Clays are considered relatively impervious, while sands and gravels are considered pervious.

Hydraulic gradient: Between two points in a hydraulic flow: the difference in total head (piezometric levels) divided by the length of the flow path (distance between the two points).

Hydrometer test: Laboratory test used to determine the amount and distribution of finer particles of a soil sample. Fine soils are classified as silts and clays. (Soils that pass the No. 200 sieve, or 0.075 mm).

Hydrostatic pore pressure: Pore water pressures exerted under conditions of no groundwater flow where the magnitude of pore pressures increase linearly with depth below the ground surface.

Igneous: Rocks that were once molten. Could be intrusive (ex. granite), or extrusive (ex. basalt).

Illite: A type of clay structure where potassium ions hold together a silica(x2) and aluminum molecule.

Immediate settlement: The settlement of a foundation occurring immediately upon loading.

Index properties: Attributes of a soil such as moisture content, void ratio, specific gravity, Atterberg limits and grain size distribution. Different from index properties.

In-situ: Undisturbed, existing field conditions.

Instrumentation: Geotechnical instruments used to monitor conditions such as deformations, pressures, loads, etc.. within the ground.

Isolated footing: Also, spread or pad footing. A footing designed to support a structural load from a single column. Usually a shallow foundation, and square or circular in shape.

Isotropic: A soil mass having essentially the same properties in all directions. Primarily refers to permeability and stress-strain characteristics.

Kaolinite: A repeated silca and aluminum molecule clay structure. White mineral. Very little volume change when drying.

Leaching: The removal of salts and soluble substances as water flows through a soil medium. Leachate is the leached liquid from the soil, rich in minerals.

Liquefaction: The sudden, large decrease of shear strength of cohesionless soil caused by collapse of the soil structure, produced by small shear strains associated with sudden but temporary increase of pore water pressure. Usually a problem in submerged poorly graded sands within the upper 50 feet of subgrade in earthquake prone environments.

Liquid limit: The water content above which the soil will flow like a liquid, but below which it will have a plastic consistency.

Liquidity index: A measure of the relationship between the current water content of a soil and its consistency limits.

Log-spiral earth pressure theory: An earth pressure theory that assumes that the failure surface is curved behind a retaining structure.

Long-term conditions: Conditions in the ground where full consolidation has taken place and the soils are fully-drained.

Machine foundation: A foundation that receives regular or irregular vibratory loads that are generated from rotating or impact machinery.

Mat foundation: Sometimes called a raft foundation. A structural slab utilized as a footing, which usually encompasses the entire building footprint. Mat foundations are advantageous on compressible soils because the building loads are distributed over a large area. Mats may also provide additional resistance to uplift.

Maximum dry density: A soil property obtained in the laboratory from a Proctor test. Density of soil at 100% compaction.

Mean normal stress: The mean value of the three orthogonal stresses.

Metamorphic: Igneous or sedimentary rock that has changed by heat and/or pressure. Examples include quartz to quartzite, shale to slate, and limestone to marble.

Micropile: Piles with relatively small diameters.

Modified Proctor: Laboratory test used to determine maximum dry density and optimum moisture content of soils. Test results are used to estimate relative compaction in the field. Soils are compacted in the laboratory in five layers with a 10 pound hammer falling 18 inches.

Modulus of subgrade reaction: Also, subgrade modulus. The ratio between the bearing pressure of a foundation and the corresponding settlement at a given point. The slope of the line in the loading range encountered by the soil in a plate bearing value test.

Mohr's circle: A circle constructed in the triaxial test using the measured stresses (principal stresses) in order to determine the stresses on the failure plane.

Moisture content: The ratio between the mass of water and the mass of soil solids. w = (wet weight - dry weight) / dry weight.

Montmorillonite: A clay structure that has a strong attraction to water. May swell 20x its original volume when saturated. A repeating molecule structure of 2 silica atoms and one aluminum.

N-Value: Also, standard penetration resistance. The number of blows required to drive a split-spoon sampler during a standard penetration test a distance of 12 inches (0.305 m) after the initial penetration of 6 inches (0.15 m).

Negative skin friction: Forces induced on deep foundations resulting from downward movement of adjacent soil relative to the foundation element.

Nominal bearing pressure: Allowable bearing pressure for spread foundations on various soil types, derived from experience which provides safety against shear failure or excessive settlements.

Normal compression line: The relationship between void ratio and the normal effective stress for soil loaded beyond the current yield stress in an isotropic compression.

Normal force: Force acting normal to the plane of reference. Units: lb, N.

Normalization: Analysis of soil test data to compensate for different states of samples.

Normally consolidated soil: Soil having a current state which lies on the normal compression line.

One-dimensional compression: Compression taking place with zero radial and horizontal strain.

One-dimensional modulus: The ratio of the change in vertical effective stress to the change in vertical strain, when there is zero horizontal strain.

Open layer: A layer or stratum of soil from which porewater may drain both upward and downward into overlying and underlying permeable layers, thus enabling two-way drainage.

Optimum moisture content: The water content at which the maximum dry density of a soil is obtained using a specific effort of compaction. Optimum moisture of a specific soil is determined through a Proctor test.

Organic soils: Earth comprised of organic material, peat, muskeg.

Overburden Soil: Overlying soil of a desirable soil stratum.

Overburden pressure: (Also overburden stress) The total or effective stress at a given depth due to the weight of overlying soil or rock.

Overconsolidated soil: A clayey soil carrying a higher load in the past. Soil having a current state which lies inside the normal compression line

Overconsolidation ratio: The ratio of maximum past pressure (preconsolidation pressure) to the current effective stress.

Overturning: Overturning failure is a result of excessive lateral earth pressures with relation to retaining wall resistance thereby causing the retaining wall system to topple or rotate (overturn).

Particle size characteristics: Used in particle size analyses and grading specifications: stated as the maximum particle size of a specified percentage (smallest) of the total content of a soil graded by mass.

D_{10} = maximum size of the smallest 10% of the particle content (also referred to as effective size)
D_{60} = maximum size of the smallest 60% of the particle content

Particle size distribution: Soil particle sizes that are determined from a representative sample of soil that is passed through a set of sieves of consecutively smaller openings.

Passive earth pressure: The maximum horizontal stress exerted by a mass of soil on a retaining surface as the surface moves toward the soil. See the link to Retaining Walls for publications, equations and calculations.

Peak shear strength: The maximum shear strength of a soil at a given normal effective stress and water content.

Permeability: A measure of continuous voids in a soil. The property which allows the flow of water through a soil. See also coefficient of permeability. See the link to Permeability for typical values relating to soil type.

pH value: A measure of acidity or alkalinity of groundwater or soil water extract based on the hydrogen ion content.

Pier: A deep foundation component that is usually cast-in-place, instead of driven, drilled or jetted as a pile.

Piezometer: An instrument used to measure in-situ pore water pressures.

Piezometric surface: An imaginary surface corresponding to the hydrostatic water level of a confined body of groundwater.

Pile: A slender member of a deep foundation system that is driven (hammered), drilled or jetted into the ground. Piles are usually constructed of timber, steel or pre-stressed reinforced concrete.

Pile Driving Analyzer: A method to compute average pile force and velocity by using wave equation analysis with electronic measurements.

Pile spacing: The distance from center to center of piles.

Piping: The movement of soil particles as a result of unbalanced seepage forces produced by percolating water. Piping leads to the development of boils or erosion channels.

Plane strain: A two-dimensional state of stress, where the out-of-plane strain (i.e. the strain normal to the plane being considered) is zero. An example of a plane strain situation would be on a cross-section through a long structure being loaded in the x-y plane, such as an embankment dam.

Plastic deformation: The distortion of soil resulting in a permanent and irrecoverable change in shape or volume.

Plastic limit: The moisture content in which a soil will have a plastic consistency.

Plastic strain: Deformation of soil that is not recovered upon unloading.

Plasticity: The property of a soil which allows it to deform continuously, usually a mass of clay size particles.

Plasticity index: The difference between the liquid limit and plastic limit of a soil mass.

Plate bearing value: A field test that is performed on compacted soil that provides an indication of shear strength pavement components.

Point-bearing capacity: Also, tip resistance or point capacity. The bearing capacity at the bottom tip of one member of a deep foundation system.

Point-bearing pile: A pile that derives the majority of its loadbearing ability from the support of the soil layer beneath the tip of the pile.

Poisson's ratio: (After Simeon Poisson 1781-1840) The ratio of the change in strain perpendicular to the direction of loading to the change in strain caused in the same direction.

Pore air pressure: The pressure of air within the void space of a partially saturated soil.

Pore pressure: Also, hydrostatic pressure. The pressure exerted by the fluid within the pores or voids in a porous material; in saturated soil the pore pressure is the pore water pressure.

Pore pressure coefficient: The ratio of the change in pore pressure to the change in deviator stress.

Pore pressure coefficient: The ratio of the change in pore pressure to the change in isotropic stress in undrained loading.

Pore pressure ratio: At a given depth of soil, the ratio of the porewater pressure to the vertical overburden pressure.

Porosity: Also, pore space. The ratio of the volume of voids to the total volume. Expressed in %, or unitless.

Pre-consolidation load: The maximum load ever imposed on a particular soil mass in its geological history.

Preconsolidation pressure: The maximum past pressure of a soil.

Pressure head: The height of a column of water required to develop a given pressure at a given point.

Pressure in tension crack: The horizontal pressure exerted in a slope or against a retaining wall due to hydrostatic water pressure in tension cracks.

Primary consolidation: The long-term consolidation of a clay from the loss of water from the voids due to a high pressure.

Principal strains: The strains occurring in the directions of the principal axes of strain.

Principal stresses: Normal stresses acting in the direction of principal axes of stress.

Proctor test: Laboratory test used to determine maximum dry density and optimum moisture content of soils. Test results are used to determine relative density in the field using in-place density tests. Standard effort or Modified effort may be used in the lab.

Quick sand: The effective stresses within a mass of sand is zero.

R - Value: The R-Value, or resistance value, of a soil is measured in a stabilometer test, and is usually estimated for pavement design. The R-value is the ability of a soil medium to resist lateral spreading due to an applied vertical load, such as tire loads. A range of values are established from 0 to 100, where 0 is the resistance of water and 100 is the resistance of steel.

Radial stress: In a triaxial sample, the total or effective stress acting perpendicular to the longitudinal axis.

Radius of influence: The outer radius from a well that has a depression of water pressure or water table produced by pumping of a well.

Raft Foundation: See mat foundation. A structural slab utilized as a footing, which usually encompasses the entire building footprint. Raft foundations are advantageous on compressible soils because the building loads are distributed over a large area. Rafts may also provide additional resistance to uplift.

Rankine earth pressure theory: Earth pressure analysis that disregards friction between the wall and soil, and assumes that failure occurs along a flat plane behind the retaining structure at an angle that is derived from the angle of internal friction.

Recompression index: The logarithmic slope recompression (reloading) line.

Relative deflection: The deflection in a foundation due to settlement.

Relative compaction: A minimum density specification usually designated as a percentage of the maximum dry density.

Relative density: Also, density index. The density of a granular soil relative to the minimum and maximum densities achieved for that particular soil.

Remolded Swell Test: Using the same equipment as the compression test, the soil sample is remolded and compacted in order to achieve similar earthwork conditions for the project's soils. The amount of swell can be measured after inundating the soil sample with water after the applied loads have stabilized.

Residual soils: Soils that have been formed in place.

Resistance value: See R-Value.

Response spectrum: Information regarding frequency-dependent energy distribution of an earthquake derived from the Fourier analysis.

Resultant force: The single force or pressure that will produce the same effect as several forces or pressures acting along a plane.

Retaining wall: Walls, usually constructed of concrete or rock, that provides lateral stability of the earth, thus preventing the soil from sloughing or slope failure. See the link to Retaining Walls for publications, equations and calculations.

Sand: Particles that pass through a #4 sieve (4.75 mm), and retained on a #200 sieve (0.075 mm).

Sand cone test: Apparatus used for measuring density of soil in the field.

Saturated density: Density of soil when the voids are filled with water.

Secondary compression index: The logarithmic slope of the secondary compression line.

Secondary consolidation: The considerably decreased rate of consolidation following the completion of primary consolidation.

Sedimentary: Soils formed by the deposition of fine-grained soil in water.

Seepage: The flow of water through soil.

Seepage force: The force transmitted to a mass of soil due to the seepage of groundwater.

Seepage pressure: The seepage force per unit volume.

Seepage velocity: The average velocity at which groundwater flows through the pores of a soil. The ratio of the volume flow rate to the average area of voids in a soil cross-section.

Sensitivity: A measure of the change in ultimate strength of clays between undisturbed and disturbed samples.

Settlement: The downward movement of soil, or the downward movement of a foundation.

Shallow Foundation: Refers to a foundation system that has a shallow founding depth relative to thc foundation width. These foundations usually include spread footings, continuous footings and mats.

Shape factors: Factors used in a general bearing capacity equation which provides an adjustment relating to the footing geometry.

Shear modulus: The ratio of the change in shear stress to the resulting change in shear strain.

Shear strain: The angular distortion or change in shape of a mass of soil.

Shear strength: The maximum shear stress which a soil can sustain under a given set of conditions. For clay, shear strength = cohesion.

For sand, shear strength = the product of effective stress and the tangent of the angle of internal friction.

Shear stress: The force per unit area acting tangentially to a given plane or surface.

Sheet pile: Steel section panels that are driven into the ground to provide lateral support.

Shrinkage index: The difference between the plastic and shrinkage limits.

Shrinkage limit: The water content corresponding to the transition between a brittle solid and a semi-solid. The water content below which further reduction in water content causes no further reduction in volume.

Sieve analysis test: Soil particle sizes that are determined from a representative sample of soil that is passed through a set of sieves of consecutively smaller openings.

Site investigation: Process of methodically observing, soil sampling and field testing in such a manner that is based on proposed development and existing site conditions.

Skin-friction capacity: Also, skin resistance or side resistance. The bearing capacity for the shaft of one member of a deep foundation system.

Skin friction stress: The shear stress on the shaft of a pile, caisson or cone penetrometer.

Sliding: Sliding failure is a result of excessive lateral earth pressures with relation to retaining wall resistance thereby causing the retaining wall system to move away (slide) from the soil it retains.

Slurry: A thick mixture of soil and water.

Soil classification: Standardized classification schemes that delineates soil characteristics that are important in determining soil behavior.

Soil Nailing: Slope stabilization method that involves installing and usually grouting closely spaced rebar in the soil or rock face.

Specific gravity: The ratio of the density of a body or a substance to the mass of an equal volume of water. Unitless.

Spread footing: Also, isolated footing. A footing designed to support a structural load from a single column. Usually a shallow foundation, and square or circular in shape.

Square Footing: Isolated/ spread footing shaped as a square. Usually a shallow footing.

Standard Penetration Test (SPT): A field test that measures resistance of the soil to the penetration of a standard split-spoon sampler that is driven 12 inches (0.3 m) with a 140-pound (63.5 kg) hammer dropped from a height of 30 inches (0.76 m). The N-value is derived from this test.

Steady state pore pressure: The pore water pressure at equilibrium when all excess pore pressures have fully dissipated.

Stiffness: Susceptibility to distortion or volume change under an applied load.

Strain: A measure of the change in size or shape of a mass of soil relative to its original size or shape.

Stress: The intensity of force per unit area; normal stress is applied perpendicularly to a surface or plane, shear stress is applied tangentially to a surface or plane.

Stress history: The past history of loading and unloading of a soil mass.

Strip footing: Also see "Continuous footing." A horizontally long footing supporting a wall. Usually a shallow footing.

Submerged Density: Also, buoyant density. Difference between the total density and the density of water. Submerged density = Soil density - Density of water.

Surcharge: An additional force applied at the exposed upper surface of a restrained soil.

Suspension: A thin mixture of soil and water, whereas the soil particles are floating within the water.

Swell: Increase in soil volume; volumetric expansion of particular soils due to changes in water content.

Swelling index: The slope of the swelling (unloading) line.

Tailings: Crushed rocks from mines after ore extraction.

Tension crack: Cracks appearing at the surface of a soil mass, often adjacent to a retaining wall or top of a failing slope.

Tension crack depth: The depth of a tension crack from the ground surface to a depth at which the horizontal effective stress is zero.

Tension pile: Piles that are designed to resist upward forces.

Time factor: A dimensionless quantity dependent on the degree of consolidation that is used in primary consolidation analyses.

Tip resistance: See point-bearing capacity.

Tolerable vibration: The level of vibration magnitude that a structure is designed. Structure vibrations range from unnoticeable to persons to structural danger.

Total head: The height of the free water surface above a given datum.

Total stress: The stress acting on or within a soil mass due to surcharges, overlying weight, etc. Total stress = effective stress + pore water pressure.

Transport soil: Soils that have formed, then moved to another place via wind, water, glacier, etc...

Triaxial shear strain: A strain parameter used in the interpretation of triaxial stress test results.

Triaxial stress test: Laboratory tests such as the *consolidated-drained (CD) test, consolidated-undrained (CU) test* and *unconsolidated-undrained (UU) test* that are used to determine the soils' strength characteristics

Ultimate bearing capacity: The bearing stress which would cause shear failure in the soil below a foundation; dependent upon the shear strength of the soil, applied loads and on the shape and depth of the foundation.

Unconfined Compressive Strength Test: Laboratory test similar to the unconsolidated-undrained test performed on plastic soils, usually clay. From this test, the *undrained shear stregth* is calculated as 1/2 of the unconfined compressive strength. Cohesion is considered to be equal to the undrained shear strength.

Undrained shear strength: The shear strength of a saturated soil at a given water content (or voids ratio, or specific volume) under loading conditions where no drainage of pore water can take place. The undrained shear strength of soil is independent of applied stresses and therefore can be measured at any level of stress, provided the void ratio remains constant. The undrained Mohr-Coulomb envelope will be horizontal.

Unified Soil Classification System: Known as USCS. A system of soil classification based on grain size, liquid limit and plasticity of soils.

Unit weight: The ratio of the total weight of soil to the total volume of a unit of soil. Units: lb/ft^3, kN/m^3.

Unit weight of water: The weight of a unit volume of water; 62.4 lb/ft^3, 9.81 kN/m^3.

Vane shear test: A field test used to measure the shear strength of a soil that is low-strength, homogeneous and cohesive.

Varved clays: Clays that are layered with fine and coarse varieties.

Vertical stress: The total or effective stress acting vertically in a soil mass at a given depth caused by the soil's own weight. See stress information in the settlement analysis and stress analysis section for information, publications, equations and calculations.

Vibrodensification: The compaction (densification) of cohesionless soils by imparting wave energy to the soil mass so as to rearrange soil particles relulting in less voids in the overall mass. Such as using a vibratory roller.

Vibroflot: A crane suspended cylindrical penetrator with an opened water jet at the tip and acts in conjunction with vibrations to dig a hole.

Vibroflotation: A method to densify granular soils using a vibroflot to dig a hole and then backfilled with sand or gravel that is dumped in from the surface and densified.

Visual Classification: A field test that is used to estimate soil characteristics such as the range of particle sizes and plasticity.

Void ratio: The ratio of the volume of voids to the volume of solids (soil grains). Unitless

Volume of solids: Volume of soil grains in a total soil volume.

Volume of voids: Volume of air space in a total soil volume.

Volume of water: Volume of water in a total soil volume.

Volumetric strain: The ratio of the change in soil volume to the original soil volume.

Water content: See Moisture content. The ratio between the mass of water and the mass of soil solids. w = (wet weight - dry weight) / dry weight.

Water table: The level in a body of soil at which the hydrostatic water pressure is zero.

Weight of soil grains: The dry weight of soil grains in a mass of soil.

Weight of water: The weight of water contained in the void space of a body of soil.

Yield point: The point at which the soil loading behavior changes from elastic to inelastic.

Yield stress: The stress at which yielding takes place in soils. The stress at which the swelling-recompression line joins the normal compression line.

Zero air voids curve: The curve created by plotting dry densities of soils corresponding to saturation at each water content.

Zoned earth dam: An earth dam embankment zoned by the systematic distribution of soil types according to their strength and permeability characteristics, usually with a centrally impervious core and shells of coarser materials.

INDEX

www.ingramcontent.com/pod-product-compliance
Lightning Source LLC
LaVergne TN
LVHW011850200926
848484LV00002B/433
* 9 7 8 1 7 8 7 1 5 3 6 4 6 *